FOREWORD BY DAVE BURGESS

THE
ENGAGING
SCIENCE
CLASSROOM

KELLY HOLLIS

Practical strategies for EVERY teacher!

MAMMOTH LEARNING

A Mammoth Learning publication
⊕ **mammoth**learning**.co**

PRAISE

An engaging read demonstrating an in-depth knowledge of the science classroom. Kelly has provided many opportunities for students and teachers alike to engage in their science classroom through practical examples of science-based experiments, but also other techniques that teachers of all levels can implement into their own classrooms. Highly recommend this book to any teacher, whether they be new or experienced.

—**Cameron Ross**, ICT Coach

Written in an engaging, easy-to-read style, this book is full of useful tips and tricks that would benefit all teachers, from those just starting out to those with years of experience. Kelly's knowledge and experience of the unique challenges of the science classroom makes this book a resource that should be on every science department's bookshelf.--Beth Kent, former head of science.

—**Beth Kent**, Educational Content and Curriculum Coordinator

Hollis strikes the perfect balance of purposeful storytelling and practical wisdom. Too often educational texts are littered with complex and nuanced terminology that makes it so difficult to understand - I felt I could absorb and appreciate everything that was being shared!

—**Jake Plaskett**, Director of Student Learning

This text provides a refreshing perspective on science teaching, offering valuable insights from the classroom teacher's point of view. Its accessible language ensures a rightful place on the professional reading list for science faculties, fostering discussions that contribute to the recognition of the science classroom as a dynamic learning environment.

—**Ken Silburn**, Science Teacher, Past President LAZSTA Met South West Science Teachers Association and 2015 Recipient Prime Minister's Award for Excellence in Secondary Science Teaching

From reading the first few lines of how Kelly started her journey to find her niche in the world, I knew this was a book I would want to read—a compendium of well-thought-out scenarios of how science can enthuse a love of lifelong learning. I have used many of them but never had the time to record them as well as has been done in this very readable, valuable contribution that should be required reading for all new teacher trainees (and seasoned veterans) educating in the Silicon Age.

—**Philip Spalding** (BSc, MSc, PGCE), STEM Teacher

DEDICATION

For my parents - for guiding me
through life's ups and downs.

For my Nan - for being my biggest cheerleader.

For my husband - for everything else.

TABLE OF CONTENTS

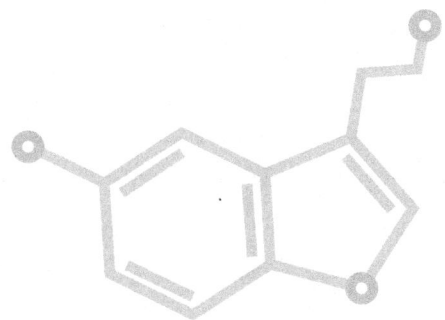

FOREWORD

Even as an avid magician, I find very little in the world more magical than science. Arthur C. Clarke said, "Any sufficiently advanced technology is indistinguishable from magic." Understanding and gaining a true appreciation for the wondrous forces that have literally shaped our past, our present, and that will, ultimately, shape our collective future, is an essential component to understanding the world. Why then has the teaching of science in our school systems so often not been filled with instructional strategies that truly capitalize on all that is amazing and wonderful in this awe-inspiring field? Rather than debate the myriad of answers to this rhetorical question and pedologicall shortcoming, let's just fast forward to the solution. Kelly Hollis, in The Engaging Science Classroom, has written a powerful manifesto for science educators everywhere in which she has tapped into years of her classroom experience bringing this subject alive for students and combined it with a curated collection of the best practices from educators all over the globe. You would be hard-pressed to read this book and not walk away with a plethora of engagement strategies that will take the lessons you teach to an all-new level that will mesmerize your students and make science a subject they truly love.

I look at teaching as a triple Venn diagram. Three interlocking circles...and we have to have all three circles to be the most powerful instructors we can be. One of those circles is content. We have to have that circle or we are just entertainers or babysitters. In this case, the content is science. I label the second circle techniques and methods. We have a whole toolbox of techniques and methods that we have accumulated over the course of our teacher training and teaching careers. Here's the problem, most educational programs, teaching books, and professional development presentations stop with those two circles. I, on the other hand, am obsessed with "The Third Circle" of presentation. Yes, you know your content. Yes, you have an array of methods at your disposal. But now, how are you going to present it in such a way that it is engaging for students? How are you going to make it relevant for them? How are you going to teach it in such a way that it draws them almost magically into your content? That's the "Third Circle" I discuss in Teach Like a Pirate, and that's the circle that Kelly expertly applies to the science curriculum in The Engaging Science Classroom. This is a one-stop shop of ideas that you can read today and literally use tomorrow.

I am a huge believer in asking, "What are students interested in outside of school and how can we use that inside of school?" Kelly effortlessly weaves together student-friendly ideas from memes, games, escape rooms, reality television shows, to entire lessons and units themed around popular culture favorites such as Harry Potter. There are ways to use what I refer to as The Safari Hook, getting students outside of the four walls of the classroom for more powerful learning opportunities. The strategies range from highly creative, reflective, and hands-on practices such as sketchnoting and journaling, all the way to the high-tech futuristic wizardry of augmented reality, virtual reality, and artificial intelligence. Do you want low-tech options? They're here! Do you want to utilize the latest and greatest in EdTech tools? She's got you! Of course, probably the best option is to combine a little bit from all of these sections to create the type of diversity in instructional strategies that constantly turns the classroom into a captivating and compelling place to learn.

Teaching is a noble profession. I am grateful for the opportunity I have had to connect with Kelly Hollis over the course of many years. Her commitment to building a professional learning community via social media started our friendship back in the early days of #aussieED. I had the great fortune to personally connect with her when I visited Australia for a couple of Teach Like a Pirate live programs. I have followed her journey ever since and have enormous respect for her commitment to education, her relentless pursuit of teaching excellence, and her true love of building strong rapport and relationships with students. Her willingness to open the vault and share with fellow educators a lifetime of engagement secrets will go a long way to empower teachers to create the kind of schools I dream of, where students are running to get in instead of out.

—*Dave Burgess*

Author: Teach Like a Pirate

President: Dave Burgess Consulting, Inc.

INTRODUCTION

Growing up, I remember having a love of science and the world around me. I was always encouraged to have an inquiring mind, and I fondly remember many holidays at my grandparents' house, exploring, creating and often setting things on fire! I don't remember having a love of learning science in my early years at school, though, with primary school and early high school science not being enjoyable for me.

Once I moved into my senior years, though, things changed. I thought I would pursue a career in forensic science (thanks to The X-Files and Gillian Anderson!), so I chose to study Chemistry and Biology for my Higher School Certificate. I enjoyed Biology and worked hard, so it was somewhat of a walk in the park, but Chemistry was a whole other ball game. I struggled to master the basic concepts in Year 11, but with the help of my teacher, I got through - just! In the end, Chemistry didn't even count as part of my university entrance score.

My first tertiary experience of science was a bit of a nightmare! First-year physics after only studying chemistry and biology was a challenge and one I had only just passed. My dream of being the next Dana Scully went out the window with my first failed subject (some crazy Maths subject that I've removed from all traces of my memory) and the massive blow to my self-esteem that came with it.

This started the rollercoaster of tertiary education for me. After abandoning the thought of solving mysteries with Mulder, I started two other university degrees and a diploma before realising I had made the wrong decision back in Year 12 when submitting my university preferences.

I had always wanted to be a teacher. Although I didn't always love my experience in the science classroom while at school, my teachers always inspired and motivated me. They always pushed me to do better, helping me achieve my goals and making my educational journey memorable. With this in mind, I walked to the education office at the university, and two and a half years later, I graduated with a double degree in science and education.

Teaching came naturally to me. I developed relationships easily with students and colleagues and always sought new ways to engage my students. Getting on to social media was the catalyst to change how I saw the world of education and the impact that I could truly have. I connected with educators across Australia and the world and began to push myself out of my comfort zone.

As I began to explore the wider world of education, I came across a book called ***'Teach Like a Pirate'*** by Dave Burgess, and it quickly became my 'edu-bible' as it was full of tangible ideas on how you can transform any classroom and dive into the deep end with your students. That book ignited my passion to become the best educator I could be, propelling me onto a trajectory I could never have imagined.

Fast forward over ten years, and here we are… my very own book. It had always been a dream of mine to write one, but I never actually thought it would happen. This book has been well over ten years in the making. I want to share with you the things that helped make my science classroom a more engaging environment for my students and helped me create memorable experiences so that when my students get 5, 10, and 15 years away from their graduation, they can look back on their science education experience in a completely different way than I did!

Disclaimer: the activities and ideas in this book have been compiled from various sources over my time as a science teacher. I do not claim to have been the original creator of each idea, merely a vessel to provide them to science teachers everywhere in one easy-to-use place.

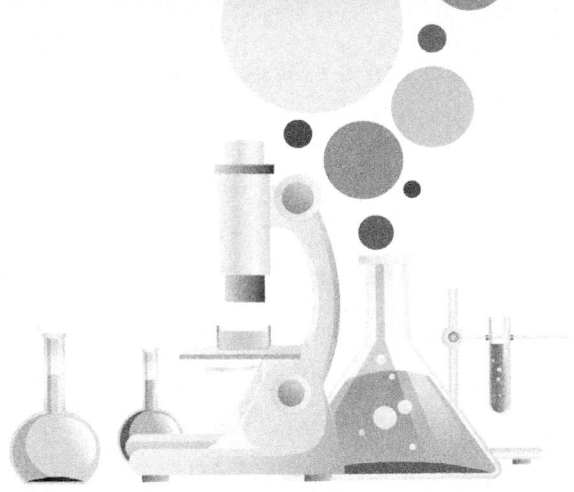

SETTING YOURSELF UP FOR SUCCESS

Before we dive into how we can create an engaging science classroom for our students, let's first explore some ways you can set yourself up for success as a science teacher. Teaching science is very different from teaching other subjects - you are often working with things that can blow up, bite you, or shock you, and I'm not just referring to the students. It can be tough to ensure you are on top of everything, so here are some ways to reduce the burden and make your teaching the awesome experience it should be.

YOUR LAB ASSISTANT IS A ROCK STAR!

They might not always be in the limelight, but trust me when I say that the 'labbies' in our schools are the real MVPs of the educational world. Without having them onside, forgetting to get your order in within the minimum time frame could mean the difference between being able to provide your students with an engaging, hands-on and inquiry-based learning experience and not!

The inspiration behind writing this book is to ensure that science is not just theory but about rolling up your sleeves and diving into the unknown. Laboratory assistants can help you set the stage for captivating experiments that turn dry concepts into memorable experiences. They are the backstage crew, making sure that the scientific show runs smoothly, materials are ready, equipment is functioning, and students have the tools to explore the wonders of science.

Safety is paramount in any laboratory setting, and laboratory assistants are the guardians of this space. Remembering every chemical and how it behaves and should be disposed of is almost impossible. Still, lab assistants meticulously maintain safety measures and protocols, instilling a culture of caution and awareness.

As science teachers, we may be the 'main act', but we can't do it alone. Laboratory assistants play the role of our trusty sidekicks, collaborating to create a rich learning experience. They meticulously set up experiments, ensuring that the laboratory is a safe, organised, and inspirational environment for our students.

Think about the last time you showed appreciation for your lab assistants - if it's been a while, make sure you thank them for everything they do for you to ensure that you can create engaging experiences for your students. I know I couldn't have done half the things I've done without

Jasmina and Ratha - for getting me through my final prac and my first full year of teaching; Denise - for helping me build confidence in the laboratory; and Cheryl, Leonie and Regina - for entertaining my crazy ideas and often helping me to clean up the aftermath - you are all the true rock stars of my classroom!

THE OFFICE STAFF ARE ALSO WORTH THEIR WEIGHT IN GOLD!

In the busy environment of a school, the office staff often work quietly behind the scenes, ensuring the seamless flow of operations. For science teachers, these office staff members are nothing short of treasures. They might not be in the classroom, but their role is immeasurable in supporting and enhancing the science teaching experience.

Science teachers often find themselves juggling a multitude of tasks, from preparing engaging lessons to managing lab equipment and materials. This is where the office staff becomes a lifeline. They can assist in managing appointments, helping you fill out the million-page excursion request paperwork and doing that last-minute photocopying you forgot. Their logistical support is worth its weight in gold, allowing you to focus on what you do best.

Beyond the paperwork and scheduling, office staff offer immeasurable emotional support. They are often the first friendly faces that students and parents encounter when they enter the school. Their warm and welcoming presence sets a positive tone for the learning environment, creating a sense of belonging and support for science teachers and their students. In a demanding profession, having a compassionate, efficient team behind the scenes can make all the difference in maintaining a positive and productive classroom atmosphere.

Finding out how your school office staff operate is always a good idea. What can they help you with to make your life easier? How can you make their life easier as well? Keeping them on your side will mean that you won't be afraid to let them know you mucked up if a disaster arises!

DON'T BE AFRAID TO MAKE MISTAKES!

On that note, it's important to remember that science teachers shouldn't be afraid to make mistakes because errors are a natural part of the scientific process. In the world of science, experimentation and discovery often involve trial and error. When teachers acknowledge their mistakes, it sets an important example for their students. It shows that making mistakes is an opportunity for growth and learning, and it encourages students to take risks and think critically. Additionally, admitting mistakes can lead to meaningful discussions and problem-solving in the classroom. Teachers and students can deepen their understanding of scientific concepts through these trial-and-error moments.

Some of my most memorable learning experiences for my students were complete disasters. Take, for example, that time I went viral! I had demonstrated the 'elephant's toothpaste' reaction many times, but this time, I didn't read my labels properly and was distracted while measuring out my reactants. The result was an explosive reaction that left a yellow stain on the roof and floor of the science lab for some time! The video made its way to YouTube, where it has over 260 thousand views and has also done the rounds on other social media outlets. Another mishap saw me accidentally burn a perfectly good $50 note while introducing the concept of combustion. Again, not paying attention left my pockets a little lighter that day - but the students definitely didn't forget, and the whole school knew within a few hours. It was certainly an expensive but memorable lesson for us all!

There's absolutely no harm in making a small mistake - but just make sure you aren't doing anything that will put yourself or your students in danger. You don't want to be the science teacher who makes their way onto the news for doing the wrong thing and almost burning the science building down!

BUILDING A PROFESSIONAL LEARNING NETWORK

It is interesting to note that more than half of the population on Earth is under the age of 30, with 96% of this demographic connected to a social networking site in some capacity. Social networking sites allow individuals to use technology to connect, collaborate and share over various platforms.

Although social networking is not a new form of communication, a rise in the use of these platforms by educators in recent times has been observed. Social networking sites allow people with similar interests to connect and build relationships through discussions around these interests. Newly created information can now be shared quickly and easily with the touch of a 'share' button, becoming instantly available for all to engage with.

Over the last decade, I have seen how social networking sites can help open the doors of communication between teachers. Since joining X (formerly known as Twitter), I have connected with many educators throughout Australia and the rest of the world. These connections (across various social networking platforms) have allowed me to associate with some of the best educators and access their shared knowledge, which has helped me transform my teaching practices.

X can be a powerful tool for creating a professional learning network (PLN). Educators can use the platform to connect with other professionals, share ideas and resources, and engage in discussions around current issues and trends in education. To create a PLN, educators can start by following other educators, researchers, and thought leaders in their field. By actively participating in the X/Twitter education community, educators can build relationships, expand their knowledge and skills, and stay up-to-date on best practices in their field.

I joined Twitter for the second time in August 2012. I had originally joined off the back of a professional development session at school that suggested we 'try this new thing'. I didn't really understand how it worked at first and didn't see how it could help me as an educator. I spent two years floating around, following a few educators here and there, but it wasn't until 2014 that the platform changed things for me. TeachMeets were gaining popularity (more on TeachMeets later!), and I signed up to present at my first one, at none other than Google, of course!

At the event, I was lucky enough to meet some of the people I had connected with on Twitter. Over dinner, Brett Salakas, Rob McTaggart, Magdalene Mattson, Kim Sutton, Nick Brieley and I began to plan a way to expand Brett's earlier work with the hashtag #aussieED and grow it exponentially, which we did. Later, we were joined by Zenia Chalich and a cluster of Melbourne-based educators. Collectively, we formed the founding #aussieED team.

#aussieED is now one of the most popular hashtags educators use in Australia and globally. Teachers use the hashtag to connect, collaborate, and share ideas about education. The goal of #aussieED was to create a community where teachers could engage in professional development and connect with like-minded individuals without needing to leave their lounge.

Since its inception in 2014, the hashtag has been used to host weekly Twitter chats, share resources, and discuss current issues and trends in education. The community also holds regular events and conferences both in-person and online. The #aussieED community is known for its passion for innovative and effective teaching practices and its commitment to improving education for all students in Australia.

Along with Twitter, Facebook groups provide another great networking opportunity and educational tool for educators. Facebook groups for educators span a range of diverse areas, such as subject-specific groups, grade-level groups, and groups focused on pedagogy and professional development. Educators can join these groups and interact with other members by sharing ideas, asking questions, and providing feedback. These groups can also be a great way to share resources with other educators, including but not limited to lesson plans, teaching strategies, and links to articles and videos.

If there isn't a Facebook group that meets the needs of a specific group of educators, you can create your own. For example, a group could be created for teachers in a particular school or region or for educators interested in a particular approach to teaching.

In 2016, while completing my Masters, I participated in a unit called 'Social Networking for Educators'. Our major project was to identify an area of need and create a social networking group to help support that need. At the time, the new NSW Biology syllabus had recently been released. As it differed from the previous syllabus, many anxious teachers sought support, resources and a place to share their concerns. Identifying this need led to the creation of the 'NSW Biology Teachers Facebook Group', which now has almost 4000 members.

NSW Biology Teachers Facebook group is a closed group, meaning only members can see the content posted in the group. The group is a platform for Biology teachers to connect, share resources and teaching ideas, ask for advice and support, and discuss issues related to biology education. Group members share information about upcoming professional development opportunities, biology conferences, and workshops. They can also collaborate to develop biology teaching resources and share tips for engaging students.

TAKE OPPORTUNITIES TO WORK ON YOUR PROFESSIONAL DEVELOPMENT

Professional learning networks are a great way to continuously work on your professional development, but for many teachers, in-person professional development can be more inspirational and motivating than online.

For science teachers, in-person professional development is like the secret ingredient in a recipe that takes your teaching to the next level. It's all about growth, staying fresh, and being the best mentor possible for your students. In a world where new discoveries and teaching methods are constantly popping up, staying on top of your game is crucial. Professional development is an essential element of your toolkit for gaining new skills, diving deeper into your subject, and being ready for any curveball the world throws at you.

It's not just about being a better teacher; it's about sparking your creativity, embracing change, and finding solutions to classroom challenges. So, whether you're learning the latest teaching tech or diving into cutting-edge research, quality professional development can be your ticket to being an amazing science teacher who can truly inspire the next generation of scientists.

TeachMeets are a great way to dip your toe into the world of professional development. Think of a TeachMeet as a science teachers' get-together, but with a twist! They're all about passionate educators coming together to swap ideas, share experiences, and inspire one another. These meet-ups offer a place where teachers can chat about their favourite teaching hacks, tech tools, or epic experiments that really hit home with students. The beauty of TeachMeets is that they're not some stuffy conference – they're like big, informal brainstorming sessions. TeachMeets are usually free and involve short and sharp presentations, no longer than 7 minutes. This fast-flowing format means you can pick up new teaching tricks, get inspired by fellow educators, and even collaborate on fun projects or research without having to sit through hour-long presentations on a topic that sounded great in the abstract but wasn't so great in person. In a nutshell, a TeachMeet is like a science teacher's dream come true!

Conferences are a more formalised version of professional development where educators converge to engage in comprehensive discussions and knowledge exchange. At conferences, knowledge is shared in longer sessions where teachers can choose the sessions they want to attend. Depending on the conference, there will be opportunities for hands-on workshop-style sessions, lecture-style sessions and panel discussions where teachers and educationalists share their teaching experiences, dissect contemporary pedagogical trends, and showcase their innovative practices. Conferences usually come with a fee and often require time off school. Most schools will have a professional development budget that will allow you to have the costs for the conference covered by the school. You can also chat with your accountant about the possibility of claiming any expenses arising from attending, such as travel and accommodation.

A great place to start exploring high-quality professional development is your state or territory's Science Teachers Association. These bodies will often hold conferences on relevant topics, produce journal magazines with up-to-date research in pedagogy and host competitions for teachers and students to take part in. Joining these associations usually comes with an annual fee, which can be paid as an individual or as a department, with membership providing discounts to events and exclusive information.

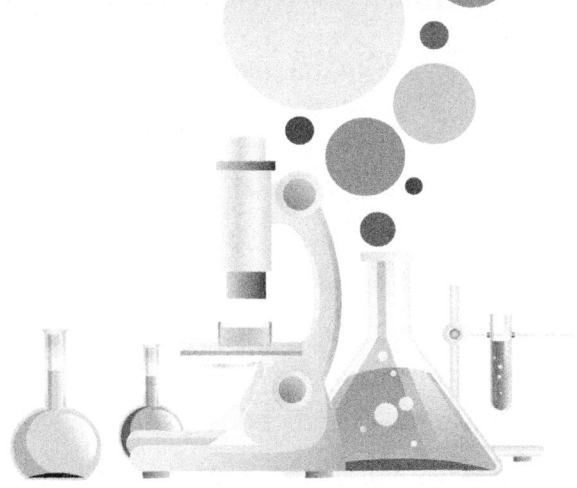

CREATING ENGAGING LEARNING ENVIRONMENTS

SETTING THE SCENE

Have you ever walked into a room and immediately felt a 'vibe' you could click with? This is what it should be like when our students walk into their classrooms. Setting the scene in the classroom is important for a variety of reasons. It helps to create a positive and welcoming environment for students, contributing to their overall well-being and success. It can also help establish clear expectations and boundaries to make the classroom more organised and structured.

One key aspect of setting the scene is ensuring the classroom's physical environment is conducive to learning. This includes things like the arrangement of furniture, the availability of necessary materials, and the overall appearance of the space. A well-organised and visually appealing classroom can help students feel more comfortable and focused, making engaging with the material easier.

Setting the scene for your students helps to establish a sense of community in the classroom. This can be achieved through activities promoting teamwork, collaboration, and open and respectful communication between students and the teacher. Building a positive and supportive classroom community can create a sense of belonging and encourage students to take risks and participate in class discussions.

Creating a safe space for your students as early as possible is one of the most important things you can do as a science teacher. It helps students feel comfortable and engaged in their learning. Science is a subject where students will be asked to move outside of their comfort zone when it comes to the content and its delivery. There are not many subjects that could be potentially life-threatening! Ensuring that your students are comfortable with you and their peers is one of the first things that needs to be established at the beginning of a school year.

As the introduction mentions, one of my favourite education books is 'Teach Like a Pirate' by Dave Burgess. In his book, Dave uses the PIRATE acronym to explore six ways teachers can boost student engagement in their classrooms. The letters of P-I-R-A-T-E stand for:

- **Passion:** whether it is content passion, professional passion or personal passion, teachers should find what they are passionate about to help pass that passion on to students.

- **Immersion:** how can teachers completely throw themselves into their lessons to engage students? Dive into the pool rather than sit on the edge!

- **Rapport:** this is one of the most important parts of the PIRATE philosophy. When teaching science, students must be comfortable taking risks and thinking outside the box. By building solid relationships with students early on, teachers can break down walls and ensure students are comfortable and willing to learn.

- **Ask and analyse:** teachers are encouraged to ask for feedback and reflect on their teaching practice, which leads to the next letter.

- **Transform:** teachers often ask for feedback but usually do not act on it! They need to use the feedback to change their practice to make the learning experiences they create better for students.

- **Enthusiasm:** none of these things can be done without having at least a small amount of enthusiasm.

Out of these six elements, the one that I focused on before all others was rapport. I made it my goal to get to know my students in the best possible way before we jumped head-first into curriculum content. This helped me find out what my students like and don't like, how they best approach their learning, and what makes them tick.

Dave shares a strategy in his book to encourage students to share something about themselves. Students use a small amount of play dough to create something that represents them and then share these creations with the class. Other students are encouraged to ask questions to learn more about their peers to build connections while allowing students to showcase their creativity.

Another way to get to know your students is to give them a blank 'Facebook' or social media template to fill out. This does not have any overly personal questions, but rather things like the student's 'favourite food/sport/TV show', how many people live in their house, and their favourite things to do outside of school. All of the answers to these questions help to build an understanding of each student and to think of ways that the things they enjoy outside of school into the classroom can be incorporated into their experience.

A space that the students feel is their learning space helps to foster curiosity and interest. When the classroom is set up with a focus on science and inquiry, it can help spark students' curiosity and interest in the subject. For example, having a classroom library stocked with science books and magazines or setting up a science corner with hands-on materials and activities can encourage students to explore and learn more about science. Ideally, This should not be static; it should change as your topics change to keep the students' curiosity alive and relevant to their learning.

When the classroom is set up to encourage a sense of community and belonging, students are more likely to feel connected to each other and the subject matter. A safe environment also promotes collaboration and teamwork amongst students and between your students and you. When the classroom is set up to facilitate collaboration and teamwork, it can encourage students to work together and share ideas. This could include seating arrangements that encourage group work, such as round tables or flexible seating options, as well as the use of group workspaces or whiteboards for brainstorming and problem-solving.

Different students have different learning styles, and setting the scene in the classroom can help accommodate these differences. For example, some students may benefit from visual aids such as diagrams and charts, while others may prefer hands-on activities or discussion-based learning. By providing various learning resources and materials, teachers can cater to the needs of all students.

When the classroom is set up in a welcoming and visually appealing way, students are more likely to feel comfortable and engaged in the lesson. This can be achieved through the use of colourful posters, interactive displays, and comfortable seating arrangements. Sometimes, you may share a classroom with other teachers, and changing things up to suit your needs may not be possible, so you need to get creative. Although not overly 'pretty', shower screens or bed sheets can be used cheaply to create portable classroom display backdrops. Items can be attached to them and folded or rolled up to be transported to other rooms. More expensive display boards that are either wheeled or carried from room to room can also be used.

Using the shared space is another way to create a welcoming and visually appealing environment for students. Hallways, open learning spaces and windows can be used to house both bought and student-created posters and displays. These are also accessible to students and teachers outside of science, so it is a great way to showcase the amazing work that students are producing or to get prospective students excited about the subject.

When teaching a unit that used a crime scene as the hook to build student engagement, I spent time after hours converting the science building foyer into a mock crime scene. When students and staff returned to school, the space had been transformed with evidence markers, a chalk outline and police tape. There were so many questions being asked by students about what it was about that when the Year 7 students arrived at class, there was already a buzz about what they would be learning in their science lessons!

In addition to creating a positive and supportive learning environment, setting the scene can help establish clear expectations and boundaries, which is especially important in the science classroom. This can be done through the use of rules and routines, as well as through explicit communication of expectations for student behaviour and participation. Clearly defined expectations can help students feel more secure and focused, making it easier for the teacher to manage the classroom and maintain a sense of order.

TEACHING SCIENCE IN A 'STANDARD' CLASSROOM

Sometimes, you may need to teach science from a 'standard' classroom due to timetabling pressures. Teaching science in a standard classroom can be a rewarding and challenging experience for both the teacher and the students. You can use this time to your advantage in many different ways.

You can use this time to get students to create rather than just consume information. For several reasons, arts and crafts can benefit the science classroom. First and foremost, they allow students to engage with scientific concepts hands-on and interactively, which can be much more engaging and memorable than simply reading about them in a textbook or listening to a lecture. When students can create their own projects, they can take ownership of their learning and feel more invested in the material.

In addition, arts and crafts activities can help students develop important skills such as problem-solving, critical thinking, and creativity. These skills are essential for success in science, as they allow students to approach challenges in a flexible and innovative way. For example, when working on a craft project, students may need to devise creative solutions to problems, such as making a model stay upright or replicating a particular shape or pattern. These types of challenges can be very similar to those encountered in scientific research, so working on craft projects can help students develop the skills they need to succeed in the field.

Science-related arts and crafts projects can effectively support dual coding learning by combining visual and tactile experiences with verbal explanations. Dual coding theory suggests that students process and remember information more effectively when it is presented in both oral and visual forms, allowing them to create mental associations between the two. This dual representation enhances learning and memory by engaging multiple cognitive pathways.

These hands-on activities enable students to create visual representations of complex scientific concepts, enhancing their understanding and retention. Students form mental images of the subject matter through creativity and imagination, reinforcing their learning. The verbalisation required during the project further connects the visual aspects with explanations, while collaborative group work fosters discussion and peer learning. By linking real-world applications to scientific principles, multisensory engagement, and personalisation of learning, these projects facilitate a comprehensive and memorable approach to science education, aligning with the principles of dual coding theory.

Some arts and crafts projects that you could carry out in a standard classroom include students creating models of cells or planets as part of a biology or astronomy unit, or they could design and build simple machines as part of a physics lesson. These types of projects can maximise the time and space you have available while also helping students better understand and remember the material they are studying, as they can see and interact with the concepts in a tangible way.

Another crafty tool that you can use with students is Foldables. Foldables are three-dimensional, interactive graphic organisers that students can create to develop their understanding of any scientific concept. First introduced to the education world by Dinah Zike over 30 years ago,

Foldables do not require any technical equipment beyond paper, pens, scissors and possibly glue. There are many templates for foldables available online that can give you inspiration to use with your students or inspire you to create your own.

Along with arts and crafts, food can be used in a number of ways to help students understand and visualise difficult scientific concepts. As good science teachers, we know that food should not be consumed in the laboratory, so these lessons are another great tool to have in your tool kit for those lessons where you aren't in a lab, or you just want to take your students outside.

Utilising food as a teaching tool in the science classroom can be a delectable way to engage students and explore various scientific concepts. For instance, in a lesson on chemical reactions, students can observe the transformation of pancake batter when it's heated on a hot plate. This hands-on activity allows them to witness the reaction where amino acids and sugars interact to create pancakes' appealing brown colour and unique flavour. In biology, dissecting fruits like apples or strawberries provides a practical opportunity to explore plant anatomy and cellular structures. Additionally, students can extract DNA from strawberries, demonstrating the molecular principles behind genetic material. Incorporating food not only stimulates students' senses but also brings science to life, making abstract concepts more relatable and memorable. It fosters curiosity and encourages students to ask questions and explore, making the classroom a more engaging and appetising place for learning.

Another benefit of teaching science in a standard classroom is allowing you to explore the topic without the distractions of a laboratory environment. If I had a dollar for every time a student walked into a lab and asked, 'Are we using Bunsen burners?' I would have a lot of dollars! Sometimes, scientific concepts need to be explored differently than in the usual experiential methods that students connect with learning science.

Science is all about asking questions and seeking answers. Encourage students to think critically about the material, ask questions, and seek answers independently. Sometimes, the best lessons can grow from providing students with the opportunity to ask questions about the topic they are studying. Simply starting the lesson with a question box can open the doors to exploring that you never thought possible. Students are great at thinking about things from different perspectives, and it is important to provide them with the time and space to ask these questions. Sometimes, you will not know the answer, but this is not necessarily bad. Seeking answers is just as important as asking questions. Setting your students the task of finding the answer to some curly questions will allow them to hone their research skills, and you'll all learn something new in the process.

TAKE IT OUTSIDE

Another way to take advantage of teaching science in a standard classroom is the ability to take your students out of the classroom altogether. By taking students into the field, they can experience firsthand the concepts they are learning about. This hands-on approach to learning can help students understand and retain information better than simply reading about it in a textbook or hearing a lecture.

When students can see the practical applications of what they are learning, it can make the material more relevant and meaningful to them. For example, visiting a local wildlife sanctuary can help students see the importance of conservation in action. In contrast, a trip to a local farm can help them understand the role of agriculture in the community. While these experiences require considerable planning, simply taking the students outside for an ecology unit and into the garden can be extremely beneficial. Outdoor science lessons help students develop a deeper appreciation for and understanding of the natural world, leading to increased environmental awareness and a desire to protect and preserve the environment.

Conducting experiments outside of the classroom can be more exciting and engaging for them than traditional classroom instruction. This can increase student motivation and participation in the lesson. Outdoor activities can also be great opportunities for students to work together and build teamwork skills. These experiences also foster a sense of community and cooperation among classmates.

Aside from the expected topics within biology that lend themselves to exploring nature and the outdoors, physics topics also provide many opportunities to get students outside and explore concepts that require more space than a conventional classroom or laboratory can provide.

When teaching a unit on motion, there are many opportunities that come up to get your students outside of the classroom. For example, students may be required to calculate the speed and acceleration of a range of objects. This can be achieved on a road near the school where students are able to observe cars and work in teams to calculate the time it takes to get from one point to the next before calculating their speed. If it is not possible to take students near a road, this can also be achieved by having students walk or run between two points, driving toy cars (either remote-controlled or pull-back) or creating paper planes and timing how long it takes them to fly a certain distance.

Newton's laws of motion can be complex concepts to understand without being able to visualise them. This is another situation where moving outside of the classroom can help. As we know, Newton's first law states that an object will stay at rest until an outside force acts upon it. But what exactly does that mean to a 15-year-old? The tablecloth trick is a common way to demonstrate this, but it can come with dangers! By taking this outside, you can get a bit crazy and try a few things to help the students remember the law. Students can also create their own version of the tablecloth trick, but with an egg, toilet rolls, a paper plate, and a cup of water… not something you'd want them to do inside on carpet or near electrical appliances.

BRINGING EXPERTS INTO THE SCIENCE CLASSROOM

Bringing experts into the science classroom can be a valuable and enriching student experience. It can provide students with real-world perspectives and insights, expose them to new ideas and ways of thinking, and help them see the relevance and importance of the subjects they are learning.

There are several ways that experts can be brought into the science classroom. One way is through guest lectures, where experts come in to give a presentation or talk on a specific topic. This can be an engaging and interactive way for students to learn from experts and ask them questions. Guest speakers are often more than willing to come and speak to your students; you just need to reach out and ask. Social media is a great tool to find experts who will come and speak with your students.

Some guest speakers that have been highly engaging in my classroom include a speaker from the Red Cross who spoke about blood donations and NRL and AFL trainers who came along to participate in a concussion in sport project. Ideally, having these people come and speak to students without needing payment is the best way to minimise the need to charge students; however, sometimes, getting a great guest speaker may involve a fee.

Another way to bring experts into the science classroom is through field trips or hands-on demonstrations. For example, students can visit a local research facility or laboratory and see firsthand the work that experts are doing. This can be particularly valuable for students who may not have the opportunity to see such facilities in person otherwise.

Online resources, such as webinars or lectures, can also bring experts into the science classroom. These resources can be accessed from anywhere and can be a convenient and cost-effective way for students to learn from experts. Again, social media is a great place to learn about these kinds of events.

One such event that my students spoke about for some time after it passed was an online question-and-answer session with Canadian astronaut Chris Hadfield. He had tweeted that he would be running this session at a time that coincided with my Year 7 Science class, so the students were all invited to submit a question. When the date rolled around, we set the classroom up in a theatre style with the live stream displayed on the data projector. We closed all the blinds and turned the lights off to create excitement and atmosphere. The students were extremely excited when 'Cara from Sydney' had her question asked as part of this international discussion. At the end of the event, students were required to create a short blog post on what they had learnt and how it had impacted them.

It is important to consider the needs and interests of the students when bringing experts into the science classroom.

It is also important to ensure that the selected experts are knowledgeable and engaging speakers who can effectively convey the information to the students.

Bringing experts into the science classroom can be a valuable and enriching experience for students and teachers. It can provide students with real-world perspectives and insights, expose them to new ideas and ways of thinking, and help them see the relevance and importance of the subjects they are learning. It can also benefit teachers, allowing them to learn from experts and stay up to date on the latest research and developments in their field.

TURNING YOUR STUDENTS INTO THE EXPERTS

One amazing way to have students showcase their understanding is for them to become experts and teach someone else about what they are learning. This can be done within the classroom by having students present in groups to the rest of the class, but a way to really make the students feel like experts is to have them present to others who are not learning about the same content as them. Or even better, to teach younger students in a way that they can understand and engage with.

If you are lucky enough to work at a school that has younger students on campus, this should be an easy scenario to create for your students. Senior students teaching Year 7 students about chemical reactions or ecosystem dynamics can be a simple way of taking advantage of your environment to support your students' learning.

If you don't have this luxury, or you want your students to engage with even younger students, one way you can have your students connect remotely with others is through the use of technology tools such as Google Meet, Zoom or Microsoft Teams. Each of these platforms allows for video conferencing to occur across a number of locations in real-time with the option of text-based chat and session recording features.

Leading up to a 'live' event, students should be encouraged to research and learn about the concepts they will teach to create age-appropriate presentations for their audience. Students can create mini-lessons that include an introduction, body and conclusion and a formative assessment activity that can be run in real-time or provided to the teacher of their 'class' to do after the session. This process is collaborative and requires the students to use critical thinking skills to determine what will be 'just right' for their students.

During the lesson, the students will take on the role of the teacher. They will share their knowledge, carry out demonstrations, and ask questions to gauge their students' understanding. Depending on how many students they are presenting to, providing opportunities for question time allows for deeper engagement from the students listening and learning and helps to demonstrate the presenting students' understanding of the content.

It is always important with all activities such as this that require a large amount of work and effort from the students to ensure that there is an opportunity for reflection and feedback. As a teacher, it is vital to provide students with constructive criticism if areas of their presentation could have been better or with praise if not. Feedback from your students and the students they presented to will also be a valuable tool to determine whether these activities are valuable in the grand scheme of your programs.

Citizen science projects also offer a remarkable opportunity to engage students in science and empower them to become experts in specific areas. These initiatives allow students to participate actively in authentic scientific research, transcending traditional classroom settings. By involving students in real-world data collection and analysis, they gain a deeper understanding of scientific concepts and methodologies while fostering a sense of ownership and responsibility in their learning journey. Students often develop a passion for a particular scientific field, transforming them into experts in their chosen areas. By becoming citizen scientists, students contribute valuable data and insights to global research efforts, which helps them feel that their learning is purposeful and has meaning..

However, both positives and negatives are associated with incorporating citizen science projects into school settings. On the positive side, these projects foster a sense of curiosity and enthusiasm for science, making learning more engaging and relevant. They promote critical thinking, problem-solving, and data analysis skills, which are crucial in the 21st century. Moreover, students learn about collaboration, teamwork, and the importance of community involvement. On the negative side, some challenges may include the need for specialised equipment, time constraints, and the potential for data quality issues. Careful planning, teacher guidance, and appropriate project selection can address these challenges and ensure that the benefits of citizen science projects far outweigh any potential drawbacks. These projects are a powerful tool for nurturing a new generation of passionate, skilled, and engaged young scientists.

You can use a Google search to find citizen science projects in your local area that may engage and inspire your students.

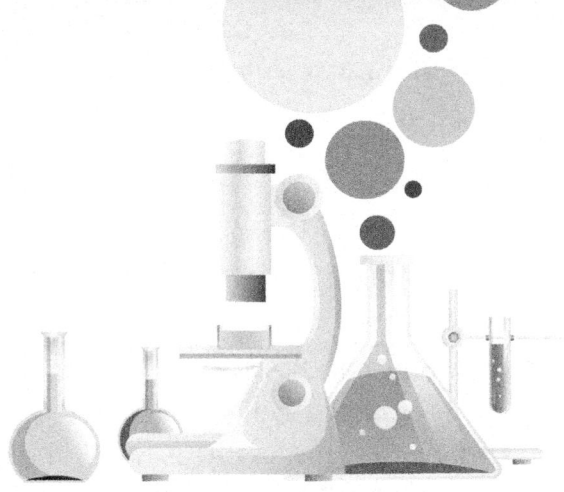

CREATING ENGAGING SCIENCE UNITS

CONTEXTUAL UNITS OF WORK

Teaching biology, chemistry, or physics is definitely fun, but do you know what is more fun than just teaching a standalone unit of work? Finding a context that you can weave into teaching and learning programs so that they are more engaging and relatable for the students. Contextual units of work are designed to intertwine elements of the students' real world into their learning to help them retain the information by being able to make meaningful connections. Contextual units can be fun or can also provide students with opportunities to explore some of the more complex parts of their world. Creating contextual units of work can also be a lot of fun for you as a teacher to design as well.

There are several benefits to using contextual units in science education. One of the obvious benefits is that it helps students develop a deeper understanding of the concepts they are learning. When students can see how the concepts they are learning apply in real-world situations, they can make connections between abstract ideas and the tangible world around them. This helps them build a stronger foundation for their learning, leading to better retention and comprehension of the material.

Another benefit of contextual units is that they help to make science more relevant and engaging for students. Many students may find traditional science lessons dry or boring, but when the concepts are presented within a real-world context, they are more likely to be interested and engaged in the material. This can be especially important for students who may not have a natural inclination towards science, as it can help to make the subject more accessible and relatable.

Contextual units can also help students develop critical thinking and problem-solving skills. Placing scientific concepts within a real-world context allows students to apply their knowledge to real-world problems and challenges. This helps them develop the skills they need to think critically and creatively, which are essential for success in any field.

Using TV and movies to teach science

Science teachers can draw a lot of inspiration from pop culture to create contextual units of work. Using pop culture references in science education can effectively engage students and make complex concepts more relatable and accessible.

Although the first Harry Potter book was released in 1997, it still provides a wealth of opportunities for teachers to explore a range of scientific concepts in a fun and engaging way. The best way to incorporate this kind of context into your classroom is to start by selecting key scientific concepts and matching them with relevant themes from the wizarding world. Consider teaching chemistry through a 'Potions Class' or using astronomy and divination to explore celestial concepts. The use of props and visual aids from the series can enhance the immersive experience. This approach makes science relatable and enjoyable and encourages students to explore the intersection of fantasy and science, sparking their curiosity and imagination.

'The Chamber of Secrets' is the name given to a unit of work that was written for Year 7 students to introduce them to the high school science laboratory. Just like the start of the Harry Potter series, students were greeted in their first lesson with a letter to inform them that they had been accepted into the 'School of Witchcraft and Wizardry'. From here, students were told that they would learn how to be the greatest wizard they could be – if the word wizard actually means scientist! Within the 'Chamber of Secrets', students were introduced to things they may never have seen before, pushed to limits they had never been pushed to before and amazed by experiences they had never experienced. The four walls of the 'Chamber' held within them a wealth of knowledge beyond their wildest imagination.

This unit was then followed by 'Spells and Potions', a unit exploring the particles of matter, but with a twist. The unit's outline stated, 'Everything around us is made of tiny, magic particles. These magic particles behave differently in different substances depending on whether the substance is a solid, liquid, or gas. These magic little particles constantly move and bump into each other, even if they are so small that we cannot see them. We can change how these magic particles behave by adding or removing energy. This makes them move faster or slower. We can even change these magic particles completely by conducting a chemical reaction. Forces can also act on these magic particles. Forces can include pushes, pulls, and twists. Magnets and electrostatic forces are invisible forces that can cause substances to do different things.'

Throughout both of these units of work, the themes, characters, and settings seen throughout the Harry Potter books and films were intertwined throughout the lessons. Scenes from the films were used as hooks at the start of lessons to get the students thinking about the concepts to be explored, while 'Educational Decrees' replaced the usual classroom rules. Students were fully immersed in the magic of science from the minute they walked into their first lesson on the topic right through until the last.

One of my favourite parts of teaching the Harry Potter units of work was dressing my classroom up like Hogwarts, or 'Hobwarts', as I called it the first time I did this (my maiden name is Hobson!). To achieve this, I purchased a brick backdrop that I used to cover the door with a sign for Platform 9 ¾, which allowed the students to cross the threshold from their everyday school corridor into the world of magic (and science!). The classical soundtrack was playing, (fake) candles were lit as students entered the room, and students were greeted by me in my full Gryffindor robes. The lesson continued with the theme of Harry Potter by exploring a range of 'magic spells' that the students could conduct based on their newfound skills. These included things like secret messages, fire writing and using plants as chemical indicators.

Believe it or not, reality television shows can also be transformed into engaging science units of work. I'm not talking about 'Keeping up with the Kardashians', but I'm sure if you tried really hard, you could make it work! Two shows that have been turned into successful science units of work by colleagues of mine include, 'The Amazing Race' and 'Masterchef'.

'The Amazing Race' can be used as the context for an otherwise dry geology unit that explores the structure of the Earth, rock types and the rock cycle before launching into space to explore the Sun and Moon and how they impact us on Earth. Like in the TV show, students are given challenges they need to complete to get the clue for their next destination. Each destination they were required to visit had a unique geological feature for them to explore. As students travelled from each location to the next, they charted their journey out on a world map and included vital information as a summary alongside each stop.

The 'Masterchef' concept can be used to create excitement and engagement around a chemistry unit of work that explores chemical reactions, including types of reactions and rates of reactions. Rather than simply having students carry out a range of chemical reactions to explore the changes that occur in the lab, the context of the kitchen and cooking was used so that students could see how important chemistry is in the creation of flavours and smells for us to enjoy the meals that we eat.

Using other relevant contexts to engage students

The 'Science of Toys' is another unit of work for students that can be used to explore how physics is required to create an engaging toy for children. Students take on the role of toy designers, and in preparation for their new job, they are 'trained' to analyse how different toys work – what makes them move, twist, turn, and fly. To do this, students learn about a range of forces and energy transformations, with the final assessment of the unit being a toy designed by the students showcasing a number of these forces and energy transformations. To further demonstrate their understanding of these scientific concepts, students also create a blog following their journey from designing to creating their toy and an advertisement that teaches 'consumers' about it.

Another context that works well to appeal to students, particularly female students, is a unit titled 'The Science of Style', which explores chemistry in the beauty industry. This unit was always very popular in the all-girls school I was teaching in at the time. The girls enjoyed learning about the cosmetics they use in their everyday lives and the ability to create their own bath bombs and lip gloss. Along with its relevant and engaging context, this unit also included various areas of science rather than simply focusing on one strand. This unit of work covered the chemistry topics of elements, compounds and their properties, and topics from physics, including energy transformations and changes.

The assessment for this unit of work can involve a group and individual component, where students carry out a makeover for 'Cinderella'. Students research different cosmetics, appliances and fabrics chosen as part of their makeover and present the chemistry and physics behind their creations. As the final group product, students showcase their makeover by creating an interactive presentation with video, images and text (visual or voiceovers), sharing the knowledge gathered.

Overall, contextual units are an important aspect of teaching science because they provide a framework for students to better understand and apply the concepts they are learning. By placing the scientific concepts within a real-world context, students can make connections between what they are learning and the world around them, which can help to make the material more relevant, engaging, and meaningful. By incorporating contextual units into their science classrooms, teachers can help students develop a deeper understanding of the material and develop students' critical thinking and problem-solving skills.

INTERDISCIPLINARY UNITS

Teaching interdisciplinary topics in science can be a challenging but rewarding task for teachers. Interdisciplinary topics involve integrating concepts and ideas from multiple scientific disciplines, such as biology, chemistry, physics, and earth science or with other subjects. These topics provide students with a more holistic understanding of the world and can also help foster critical thinking skills. Creating interdisciplinary units of work in science is a powerful educational approach that fosters the development of transferrable skills in students. Students are encouraged to think critically, problem-solve, and apply their knowledge across various contexts by seamlessly blending science with other subjects like mathematics, technology, or the arts.

These interdisciplinary experiences deepen students' understanding of scientific concepts and nurture skills such as teamwork, communication, and adaptability, which are invaluable in a wide range of contexts. Furthermore, students gain a holistic perspective, recognising the interconnectedness of different fields, which is crucial in addressing complex, real-world challenges. In essence, interdisciplinary science units empower students to become versatile learners equipped with the tools they need to succeed in an ever-evolving, interdisciplinary world.

One approach to teaching interdisciplinary topics starts by identifying the key concepts students need to understand to grasp the topic. For example, if teaching about climate change, you should cover concepts such as the greenhouse effect, carbon dioxide emissions, and the role of the oceans in regulating the Earth's temperature. Breaking down the topic into smaller, more manageable chunks helps students to better understand the interrelationships between different scientific disciplines and how they contribute to a larger understanding of the topic.

Another important aspect of teaching interdisciplinary topics is using real-world examples and case studies to help students see how they can apply the concepts they are learning. For example, you might use examples of how climate change affects different ecosystems or how different technologies are used to address environmental issues. By relating the concepts to authentic situations, you can help students better understand the topic's importance and how it relates to their own lives.

In addition to using real-world examples, it can also be helpful to incorporate hands-on activities and experiments into your lesson plans. This helps students gain a deeper understanding of the concepts being taught and helps to make the subject more engaging and interactive. For example, you might have students conduct experiments to measure the greenhouse effect or to test the acidity of different bodies of water.

Creating a positive and supportive learning environment when teaching interdisciplinary topics is also important.

This can involve encouraging student participation and discussion, providing feedback and support to help students understand difficult concepts, and creating a safe and inclusive space where all students feel welcome and valued. Creating a positive learning environment can help students feel more motivated and engaged in the subject matter.

It is important to be flexible and open to trying new approaches when teaching interdisciplinary topics. Each class and group of students is unique, and what works for one group may not work for another. By being open to trying different approaches and adapting your teaching style to meet the needs of your students, you can ensure that your lessons are effective and engaging.

Collaborating with teachers from different subjects to create interdisciplinary units of work can be a highly rewarding but occasionally challenging endeavour. On the positive side, working across subject areas cultivates a dynamic environment where educators can exchange expertise, teaching methods, and fresh perspectives, ultimately enhancing each individual's professional development. This cooperative approach also ignites creativity and innovation as teachers combine their unique skills to craft engaging, multifaceted lessons that captivate students.

In the classroom, students benefit from exposure to various teaching styles and are encouraged to think critically, solve problems, and communicate effectively across different knowledge domains. Furthermore, it fosters a sense of community within the teaching staff, fostering a supportive network where educators can draw inspiration from one another, leading to a richer and more comprehensive educational experience. While challenges may arise, the shared benefits of this collaborative effort make it a rewarding and transformative experience.

Challenges may arise in coordinating schedules, aligning curriculum goals, and addressing differences in teaching styles and priorities. To overcome these obstacles, open communication and a shared vision are essential. Regular meetings and a clear project plan ensure everyone is on the same page. Flexibility is also key, allowing individual teachers to adapt their content to fit the interdisciplinary framework. Ultimately, the advantages of interdisciplinary collaboration, in terms of enriched education and transferable skills, make the effort to overcome these challenges well worth it.

PBL IN SCIENCE

Project-based learning (PBL) is a teaching approach that involves students actively investigating and responding to authentic, complex problems or challenges. In a PBL science classroom, students engage in hands-on, experiential learning activities that allow them to apply their scientific knowledge and skills to solve problems or answer questions related to a particular topic or theme.

There are several benefits to using PBL in the science classroom. First, PBL promotes active learning and critical thinking. Instead of simply being told information, students can discover and explore concepts independently, which helps them better understand and remember the material. PBL also encourages students to think creatively and to consider multiple perspectives, which can help them to develop problem-solving and decision-making skills.

PBL is well-suited for science education because it allows students to engage in authentic, real-world problem-solving. By working on relevant and meaningful projects, students can connect their learning to their own experiences and interests, increasing their motivation and engagement. PBL also allows students to apply their scientific knowledge and skills to authentic problems, which can help them better understand the relevance and practical applications.

The incorporation of PBL in the science classroom can foster collaboration and teamwork. In a PBL environment, students often work in small groups or teams to complete projects, which allows them to learn from one another and develop important social and communication skills. Working on projects together promotes community and encourages students to share their ideas and knowledge.

PBL projects also offer the opportunity to use a range of activities and resources, allowing them to be tailored to individual students' needs and interests and ensuring that all students can engage in the learning process and find success.

Finally, PBL can be a highly effective way to integrate technology and other 21st-century skills into the science classroom. Using technology and other resources, students can research, analyse, and present their findings in various formats, which can help them develop important digital literacy skills.

I have been lucky enough to work with two of the best PBL educators in Australia, Jake Plaskett and Celinda Corsini. I learned a wealth of information from them about the best ways to integrate PBL into the science classroom. During the annual 'Festival of Creative Spirit', a week-long program where students choose a project based on their own interest from a range of projects pitched by teachers, I ran a unit titled 'Your Brain On Sport'.

The project pitch delivered to students included an overview of the problem, which is the increased rates of concussions in professional sports. The students were encouraged to think about the short and long-term effects of concussion and what various sporting codes do to reduce these for players. Students were also encouraged to look at the construction and design of headgear worn by players and investigate whether they effectively protect the brain. As part of the week, several medical professionals who are experts in the field of concussion, as well as sporting players directly affected by concussion, were invited to speak with the students to share their experiences and knowledge.

This project allowed me to combine my love of rugby league with my love of science to help engage the students who chose our project. Before the official beginning of the project, the students who elected to participate were invited to attend a live game at Leichhardt Oval (the 8th wonder of the world!) between the Wests Tigers and the North Queensland Cowboys. This allowed them to experience the game firsthand, observe how players interact, see how some wear headgear and better understand the impacts of tackles, thus developing a strong prior knowledge before starting the unit

Before the project, I had also reached out to a number of NRL personalities on social media to ask about the possibility of representatives from the NRL to come and speak with the students. The response was amazing, and as a result of this call out, I was able to organise one of the trainers from the Australian Kangaroos side to come along and run the students through the concussion protocols that happen at games as well as the other things the game is doing to try to minimise the effects. We were also lucky enough to have a trainer from the Greater Western Sydney (GWS) Giants AFL team come and speak with the students to give the same information from an AFL perspective.

The week also involved a lot of investigations - exploring Newton's laws of motion and how they relate to concussion, materials that can be used to cushion the impact of concussion and the changes to the brain that occur when a player is concussed. Students designed 'helmets' for eggs to protect them from a 3-metre drop and changed their designs to get the lightest and most effective protection.

The project culminated in a showcase for teachers and parents where students displayed the newly designed headgear they had built from recycled materials or printed with 3D printers. This required students to develop new skills, work collaboratively, and continuously rethink their ideas. They also needed to be as confident with their materials as possible, as they were required to talk with parents about what they had learned and the project they had completed.

Another example of a PBL unit for an ecology topic is titled 'Animal Re-Zoo-Dence', where students must redesign a zoo enclosure for an animal of their choice. This STEM project includes input from the students' science, maths, and technology teachers throughout the term as the students develop the required skills to create their enclosure.

The PBL Entry Event for this unit involves an excursion to the zoo. During the excursion, students are encouraged to spend as much time as possible at the enclosure for their chosen animal to take in as much information as possible about the current design. As part of their observations, students make note of things like the different behaviours the animal displays in the different parts of the enclosure, the different ways that the public can interact with the animals based on the enclosure design, and the ease of access for the zookeepers to do their job of looking after the animal.

Upon returning to school, the students learn different concepts required to support them in their final project design in their respective classes. In science, students will spend time researching their animal of choice regarding their habitat, food choices, and adaptations. Students engage with this content as part of a self-directed unit of work, allowing them to progress at their own pace. This learning mode encourages students to work independently on different tasks, and they can showcase their learnings on a website they design.

A group element of the project involves the students creating a scaled model of their re-designed enclosure, with each enclosure including at least one element created using a 3D or laser printer. Students also create the remainder of their enclosure using recycled materials and complete it within class time. Lastly, students present their finished product to their parents and peers at the end of the project showcase.

Project-based learning is a highly effective teaching approach for the science classroom because it promotes active learning, critical thinking, real-world problem-solving, collaboration, personalisation, and the integration of technology and 21st-century skills. By engaging students in hands-on, experiential learning activities, PBL can deepen their understanding of scientific concepts and develop important scientific and life skills.

COMPETITION IN THE SCIENCE CLASSROOM

Competition can be a powerful motivator in the science classroom. You just need to watch any playground of primary school or even younger students, and you will see that competition is a natural part of their development. Children spontaneously seek competition with their peers and seem to have this innate desire to compare themselves to others in every way. From here, it is a small step to physical and intellectual competitions where children compete at a higher level in sports through individual and team activities.

Along with the innate desire to compete, there is the natural desire to learn, so it is not surprising that competition and education can be related and used together.

Well-organised competitions and educational challenges can drive students to give their best and enhance motivation and learning. It is important to use competition to promote learning and collaboration rather than simply promoting winning at all costs. Fun competitions can be inclusive, allowing all students to experience success and, as a result, boost their confidence. A good competition should challenge all participants to give their best.

Along with this, there are many other benefits that competition can bring for students when done well:

- Competitions create a framework for practising a growth mindset; students can look for ways to improve themselves and, as a result, their place within the competition.

- Students are often required to think on their feet and determine the next best course of action.

- Students develop the skill of self-motivation and develop a strong sense of agency, especially in individual competitions.

- Competitions allow students to frequently assess the risks and evaluate their learning progress, which is often missed in traditional teaching.

- Students learn that 'losing' does not have to impact their self-worth. Teachers can work with students to build resilience in handling these situations. Students also have the opportunity to learn sportsmanship and win gracefully, both important life skills beyond the classroom.

One way to incorporate some friendly competition and external motivation into the classroom is to organise science quizzes or trivia games that students can play in teams. These games can be designed to cover the topics that the students have learned in class, and they can be a fun way for students to review their knowledge and test their understanding. Teachers can even offer rewards for the winning team, such as extra credit or a small prize.

Another way that teachers can use competition in the science classroom is by setting up science fair competitions. Students can be tasked with designing and carrying out their own scientific experiments, and they can present their findings to their classmates and judges. This type of competition can be a great way for students to learn about the scientific method, develop their critical thinking skills, and practise their communication skills. Teachers can offer rewards for the best experiments, such as certificates or recognition, in front of the whole class.

Teachers can also use competition in the science classroom by setting up challenges that require students to work together in teams to solve problems. For example, teachers can ask students to design and build a bridge using only popsicle sticks and glue that can support a certain amount of weight. This type of challenge can help students develop their problem-solving skills, encourage teamwork and collaboration, and promote creativity and innovation. Teachers can offer rewards to the most successful teams in their challenges, such as bragging rights or a class celebration.

If you are looking at bringing the element of competition into your classroom, here are some tips to consider:

- Make it about learning: competition should be used as a way to encourage students to learn and improve, rather than just trying to beat their classmates. Set clear learning goals and criteria for success, and use competition to encourage students to meet those goals.

- Use a variety of competitions: there are many different ways to incorporate competition into the science classroom, including science fairs, quizzes, and challenges. Using a variety of competition formats helps keep students engaged and motivated.

- Encourage collaboration: while competition can be a helpful motivator, it's also important to encourage students to work together and collaborate. Encourage students to share ideas and work together to solve problems rather than simply trying to outdo each other.

- Set clear rules: establishing clear rules and guidelines for competition can help to ensure that it is fair and promotes learning. Make sure that students understand the rules and what is expected of them.

- Use it sparingly: competition can be a useful tool, but it's important not to rely on it too heavily. Use it as one of many different strategies for engaging and motivating students rather than the primary method for teaching science.

- Recognise effort and improvement: it's important to recognise and reward students for their efforts and improvements rather than just focusing on the winners. This can create a positive and supportive classroom culture.

Along with developing competitions within your class for students to participate in, there are also external competitions you can enter your students into. Many of these kinds of competitions are available, and a quick Google search will help you find relevant competitions for your students in your region.

The **Brain Bee Challenge** (thebrainbee.org) is an international competition that offers students a unique opportunity to learn about the brain and its functions, neuroscience research, and careers in neuroscience and to dispel misconceptions about neurological and mental illnesses. In Australia and New Zealand, the Brain Bee Challenge is run across a number of rounds. The first round is open to all Year 10 (Australia) or 11 (New Zealand) students. It involves spending time engaging with neuroscience-based learning material on the Education Perfect platform before sitting an online assessment to determine who moves through to Round 2. Round 2 involves a regional in-person event where students collaborate on various activities and ends with a second assessment of the student's understanding. The third round sees the regional winners come together for the national finals. The winner from each national final is then invited to the International Brain Bee Challenge event, which coincides with a large international neuroscience conference. Other regions may conduct the challenge slightly differently; however, all culminate with the international final.

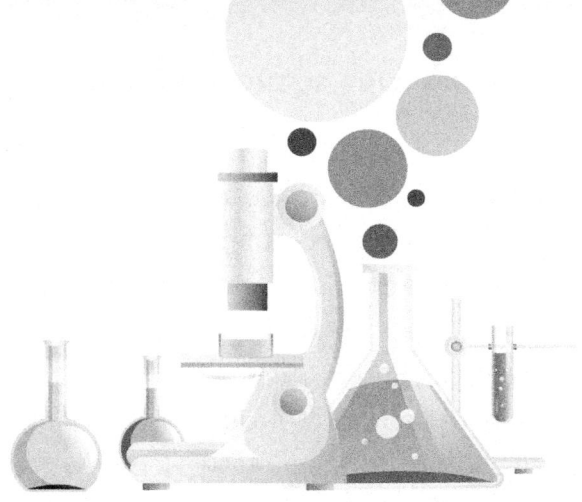

CREATING ENGAGING OPPORTUNITIES TO IMPROVE LEARNING OUTCOMES

LITERACY AND SCIENTIFIC LITERACY - WHAT'S THE DIFFERENCE?

As science teachers, we are responsible for ensuring that our students are proficient in scientific concepts and principles and capable of effectively communicating their understanding. This requires a fundamental understanding of literacy and scientific literacy.

Literacy refers to the ability to read, write, and communicate effectively using language. It is a crucial life skill that enables individuals to comprehend and articulate their thoughts and ideas clearly. It also plays a vital role in our daily lives, from reading street signs to filling out job applications. On the other hand, scientific literacy refers to understanding and using scientific concepts and principles to explain the natural world. It involves the application of scientific thinking and reasoning to solve problems and make informed decisions. Scientific literacy is not just about memorising facts but understanding how scientific knowledge is generated and how it can be applied in various contexts.

While literacy and scientific literacy may seem like distinct skills, they are deeply intertwined. Effective scientific communication requires understanding scientific concepts and the ability to articulate ideas clearly in writing and speech. Conversely, developing literacy skills can help students better comprehend and communicate scientific concepts.

Moreover, scientific literacy is becoming increasingly essential in our modern world. From climate change to public health, scientific issues are at the forefront of many societal debates. Citizens must be scientifically literate to make informed decisions and actively participate in these discussions.

In conclusion, literacy and scientific literacy are essential skills students must develop to succeed academically and professionally. As science teachers, we must strive to provide our students with a robust education in both areas, emphasising the importance of clear communication, critical thinking, and effective argumentation. By doing so, we can equip our students with the tools they need to become scientifically literate individuals capable of understanding and contributing to the scientific discourse.

ENGAGING LITERACY-BUILDING ACTIVITIES

Sketchnoting

Sketchnoting is a visual note-taking technique that combines images, symbols, and text to represent ideas and concepts. Sketchnoting is a powerful technique in science education that links seamlessly with the concept of dual coding. Dual coding theory suggests that students are more likely to remember and understand the content when they engage with information using both visual and verbal channels.

Sketchnoting takes advantage of this by encouraging students to create visual representations of scientific concepts alongside written explanations. This practice helps students visualise complex ideas and reinforces their understanding by translating abstract concepts into tangible images. By combining words and images in their notes, students employ both verbal and visual memory pathways, deepening their comprehension and retention of scientific knowledge.

Additionally, sketchnoting can be a helpful way for students to engage with their notes and stay focused during class or when creating revision notes. Using sketch notes, students can transform their notes into engaging and visually appealing study aids.

In the science classroom, sketchnoting can be used as a tool for students to take notes on lectures, readings, and discussions. Rather than simply writing down text-based notes, students can use images and symbols to represent scientific concepts, making their notes more engaging and memorable. Additionally, sketchnoting can be a helpful tool for students to use when working on group projects or when presenting their research findings. By creating sketches of their ideas or findings, students can create a visually engaging presentation that will capture the attention of their classmates.

Sketchnoting can be used to help students to develop their creativity and critical thinking skills. By encouraging students to use their imaginations to represent scientific concepts through visual elements, sketchnoting promotes creative thinking. Additionally, sketchnoting requires students to distil complex scientific concepts into simple, easy-to-understand visual representations, which can help them develop critical thinking skills.

Journalling

Journalling is a reflective writing practice that can be used to encourage students to think deeply about their experiences and ideas. In the science classroom, journalling can be used as a tool for students to reflect on their learning and to document their progress as they work on scientific projects or experiments. Journalling can also help students make connections between scientific concepts and their own lives, promoting deeper understanding and engagement with the material. Students can develop their writing skills and become more reflective, critical thinkers by engaging in regular journalling.

In the science classroom, journalling can take many forms. Students can keep a traditional written journal, recording their thoughts, observations, and questions related to scientific concepts. Alternatively, students can use digital journals or blogs to document their learning and share their ideas with classmates. Teachers can provide prompts or guiding questions to encourage students to reflect on specific aspects of their scientific learning, such as the process of conducting experiments or the ethical implications of scientific research. Additionally, journalling can be used as a tool for self-assessment, allowing students to reflect on their progress and identify areas for improvement.

Finally, journalling provides an opportunity to promote interdisciplinary learning in the science classroom. By encouraging students to connect scientific concepts and other subject areas, journaling can help students develop a more holistic understanding of the world around them. Overall, journalling is a versatile and effective tool that can be used in various ways to help students engage with scientific concepts and develop important skills.

Writing a Letter to the Editor

Writing a letter to the editor is an excellent activity that can support the development of students' literacy and scientific literacy skills. Writing a letter to the editor requires students to engage in critical reading and analysis of a text, identify the key arguments and issues, and develop their viewpoints. It also requires students to communicate their ideas effectively and persuasively in writing, an essential skill in many academic and professional fields.

One example of a topic where students could write a letter to the editor is climate change. Climate change is a complex issue that requires understanding scientific concepts such as carbon emissions, greenhouse gases, and global warming. Writing a letter to the editor on this topic would require students to research and analyse the causes and effects of climate change, evaluate the effectiveness of different policy proposals, and articulate their own opinions on the issue.

Another example of a topic for students to consider when writing a letter to the editor is the issue of animals in captivity. This controversial issue requires students to research and analyse the ethical considerations surrounding keeping animals in zoos and aquariums. Writing a letter to the editor on this topic would require students to critically evaluate the impact of captivity on animals' physical and psychological well-being, weigh the benefits and drawbacks of zoos and aquariums, and express their perspectives on whether or not keeping animals in captivity is morally justifiable. This activity would improve their literacy and scientific literacy skills, help them develop empathy and compassion towards animals and cultivate a sense of responsibility towards the natural world. This particular activity could be included in the 'Animal Re-Zoo-Dence' PBL unit mentioned earlier.

ENGAGING GAMES AND ACTIVITIES TO BUILD SCIENTIFIC VOCABULARY

Games and activities can play a pivotal role in supporting the development of specific vocabulary in the science classroom. By making learning enjoyable and interactive, these methods help students engage with scientific terms and concepts more effectively. Whether through word-based games like 'Taboo' or visual activities like 'Four Pictures, One Word', students learn the vocabulary and internalise its meaning and context. Games and activities help transform the often daunting task of vocabulary acquisition into a dynamic and memorable learning experience, empowering students to communicate and comprehend scientific ideas confidently.

Four Pictures, One Word

Four Pictures, One Word is a popular game where players are presented with four seemingly unrelated images but have a common word that links them together. Players must examine the pictures closely and use their critical thinking skills to identify the common word that relates to all four pictures.

This game can be an excellent tool for building vocabulary in the science classroom. Teachers can use this game to introduce students to new scientific vocabulary terms and concepts. For example, a teacher could select four images representing different plant types, such as a cactus, sunflower, fern and palm tree. The common word that links these images is plants, which would be the vocabulary term the teacher is trying to teach the students.

You can also use this game to reinforce previously learned scientific vocabulary terms. For example, the teacher could choose four images representing different forces: gravity, friction, tension, and compression. The common word that links these images is forces, which the teacher could use to review the previously learned vocabulary terms with the students.

This game can be transferred to any topic to support students in developing vocabulary. You can also get the students involved by encouraging them to draw or provide four images relating to the key term they have chosen for their partner to guess. This game can be done with or without technology!

Odd One Out

The game **Odd One Out** involves presenting a group of related items or concepts and asking the players to identify which one does not belong or is different from the others. This game can be used as a warm-up activity or a quick assessment tool to assess students' critical thinking skills.

The Odd One Out game can be used as a vocabulary-building exercise in the science classroom. Teachers can select a group of related scientific terms or concepts and ask students to identify the odd one out. For example, a teacher might present a list of terms related to cellular biology, such as cell membrane, cytoplasm, ribosomes, and bones. In this case, the odd one out is bones, which are not a component of a cell. This exercise can help students to better understand and retain scientific vocabulary by identifying the unique characteristics of each term or concept.

Another way to use the Odd One Out game in the science classroom is to present a group of scientific concepts and ask students to identify the one that does not fit with the others. For example, a teacher might present the concepts of Newton's laws of motion, Boyle's law, Archimedes' principle, and the periodic table. In this case, the odd one out is the periodic table, which is not a law or principle related to motion and force. This exercise can help students develop their critical thinking skills and better understand how different scientific concepts are related.

Similar to the Four Pictures, One Word game, this game can be used in science classrooms that are studying any topic. Students can also be encouraged to draw or find four images around a particular topic, three related and one the 'odd one out' to combine the students' visual literacy with their critical thinking skills. Students can make this as easy or difficult as they like by thinking outside the box!

Pictionary

Pictionary is a popular drawing and guessing game where players take turns drawing a word or phrase while their teammates try to guess what it is. The player who is drawing cannot use any letters or numbers but can use gestures and body language to give clues to their teammates. The goal is to be the first team to correctly guess the word or phrase within a set time limit. Students of all ages and skill levels can enjoy this fun and interactive game.

A game of Pictionary can be modified to be played in any class. For example, in biology, students could draw diagrams of various biological processes or structures, such as the human digestive system or a plant cell. This would encourage students to think creatively and communicate scientific concepts visually. Additionally, it can be a fun way to review and reinforce scientific terminology and concepts in an engaging and interactive way.

A fun way to make this game more competitive is to have the students split into groups, and one member from the group comes to the front of the room to get their 'word'. Students then return to their group and only begin drawing when given the signal. The first group to identify the key term from the drawing is the winner for that round.

Scattergories

Another popular game that can be transformed for the classroom is **Scattergories**. In this game, players are given categories for which they need to identify words, the catch being that all words need to start with the same letter of the alphabet. This game can be modified to suit any age group and topic that is being taught. The game can be played in rounds, and each round may have the same category. However, a different letter is to be used each time a round starts.

To play Scattergories with your students, choose a category related to your current topic. Students then have a set amount of time to come up with as many words as possible that fit the category and start with the designated letter. For example, if the category is fruit and the letter is A, players might come up with apple, apricot, and avocado. Points are awarded for each unique word a player comes up with, and players can earn bonus points for creating particularly creative or unusual words.

Taboo

Taboo is a word-guessing game in which players take turns trying to get their teammates to guess a particular word without using certain 'taboo' words or phrases. Players are given a card with their target word written on it, and they have to describe the word without using any related taboo words or phrases listed on the card. The goal is to get as many correct guesses as possible within a set time limit, and the team with the most points at the end of the game wins.

Taboo can be a fun and engaging way to review and reinforce scientific concepts in the classroom. For example, in a biology class, students could be given a set of cards with biological terms and challenged to describe them without using certain taboo words, such as related scientific terms or jargon. This encourages students to think creatively and communicate scientific concepts clearly and concisely.

Similarly, in a chemistry class, students could be given cards with chemical elements or compounds and challenged to describe their properties and characteristics without using related scientific terminology. This can be a great way to review and reinforce scientific concepts in a fun and interactive way.

Spelling Roll-a-Word

Spelling Roll-a-Word is a vocabulary and literacy-building activity that can be used for any topic with students of any age. In this activity, students are broken into groups and given a list of keywords from the topic they are studying, six instructions, and a dice.

Starting with the first word in the vocabulary list, students roll a die and complete the activity for the number they roll. For example, if a student rolls a 3, they may need to 'Write a synonym of the word', or if they roll a 5, they may need to 'Write the definition of the word in language a primary school student would understand'. Once the student has finished the activity for their first vocabulary word, they continue with the rest of the list.

This activity can be differentiated to make the words or the required activities easier or harder, depending on the age of the students and the topic being studied at the time.

⚀	Write the word in a sentence.
⚁	Draw a picture of the word.
⚂	Write a synonym of the word.
⚃	Write an antonym of the word.
⚄	Write the definition of the word in language a primary school student would understand.
⚅	Write the definition of the word three times.

Celebrity Heads

Celebrity Heads, also known as Who Am I? or Guess Who? is a guessing game where players try to identify a celebrity's name that is written on a card attached to their forehead or back. Players take turns asking yes-or-no questions to try and narrow down who they are. The goal is to be the first player to guess their celebrity identity correctly.

This game can be reimagined for the science classroom by using relevant terminology instead of 'celebrities' throughout the game. Simply asking or responding requires students to recall and understand the content. Working in teams brings collaboration, a touch of competition, and team spirit. You can use the questions and responses during the game to guide student understanding, and any misconceptions can be addressed in supplemental sessions or by providing students with relevant resources.

Heads Up

Heads Up was a game introduced by Ellen Degeneres on her popular talk show. It gained mass popularity when she released an app version, which people worldwide played.

To play the game, one player holds a smartphone or tablet against their forehead with a word or phrase displayed on the screen. The other players then give clues, either with words, sounds or actions, to help the player guess the word without actually saying the word itself. If the player guesses the word correctly, they tilt their device downward to move on to the next word. If playing the game on the app, it can be played with different categories, such as movies, music, or animals, and can be customised with personalised decks.

This game can be modified for the classroom so that it does not require the use of the phone app and can include any terminology being studied at the time. In a physics class, for example, students could be given a set of cards with different forces or astronomical features on them and take turns holding up the cards while their classmates give clues to help them guess the correct answer. Similarly, in a chemistry class, students could be given cards with chemical elements or compounds and challenged to guess their properties or characteristics based on the clues given by their classmates.

Bingo

Unlike a 'normal' game of bingo where students simply cross off numbers that are called out from their sheet, students will need to match definitions to key terms in this version of bingo!

Bingo sheets should be created to prepare for this game with a range of key terms related to the concept being studied. These sheets can be laminated so that they can be reused. The teacher then stands at the front of the room and reads out a number of definitions. If the student has the matching key term to the definition being read, they cross the key term off their sheet. When the student has three terms across, down, or diagonally (or any other combination you wish!), the student calls out 'Bingo!'.

Mnemonics

Mnemonics have been used for many years to build vocabulary and memorise information. They are memory aids that help learners remember and recall information by associating it with something more memorable or easier to remember. Using mnemonics in the science classroom can help students learn and retain complex concepts and terms.

One benefit of mnemonics is that they can help students quickly recall information, which can be particularly useful during exams or when working on complex problems. Additionally, they can make learning more fun and engaging, increasing student motivation and interest in the subject matter. There are many ways to use mnemonics in the science classroom, such as creating acronyms, using imagery and visual cues, or linking information to familiar concepts or experiences.

Science educators can use mnemonics to help students understand and remember important scientific concepts and terminology better. Some extremely popular mnemonics already exist to help students remember a set of information easily.

One common mnemonic that is used is Roy G. Biv - This is a mnemonic used to remember the colours of the visible light spectrum in order: red, orange, yellow, green, blue, indigo, and violet, using the first letter of each colour to create the mnemonic. This tool is useful for students learning about light and colour in physics. Another fun mnemonic that can be used with younger students is 'My Very Excellent Mother Just Served Us Nine Pizzas' to help students remember the order of the planets (including the now-downgraded Pluto!).

Mnemonics can also be adjusted to suit your students' age and create that lasting memory. For example, a mnemonic that can be given to students to help them remember the order of classification can be pitched at younger students with 'King Phil Calls Ordinary Families Generous and Special', or you could make it a bit more exciting with the older student version, 'Keep Putting Condoms On For Good Sex'!

Collaborative crossword

A collaborative crossword involves students working in pairs to complete a crossword; however, each student only has half of the crossword in front of them.

To prepare for this activity, create two versions of a crossword puzzle made up of terminology from the current topic being studied or a range of topics to provide an opportunity for revision. With the two versions of the crossword, one will only have the down answers, and the other will have the across answers.

Divide the class into pairs and distribute Sheet A to one student and Sheet B to the other. Sheet A contains the down answers, while Sheet B contains the across answers. To complete the crossword, students must give each other clues to work out the answers and write them in the grid. When they have completed the crossword, the students are to work collaboratively to create a complete set of clues for ALL across and down words.

Four Corners Activity

In this activity, students work together to create the best possible answer to a question that is given to them. Students should be placed into groups of four, and each student should receive a sticky note large enough to write their answer on. Each student writes their best answer to the question that has been posed on their individual sticky note. Once all students have answered the question, they will place these into one of the four corners of an A3 sheet.

Collaboratively, students will use pieces from each person's sticky notes to determine the best answer to the question posed. This activity can be done for any question; however, it is a great activity when preparing students for answering examination-style questions, where teachers can provide a marking criteria to support the students in developing the best possible answer from the suggestions provided.

NUMERACY IN SCIENCE - WHY IS IT IMPORTANT?

Numeracy is understanding and using numerical concepts, operations, and relationships to solve problems and make informed decisions. Numeracy skills are essential in science, enabling students to work with data, make calculations, interpret graphs, and communicate scientific ideas effectively. Science involves students regularly collecting, analysing, and interpreting data and skills in numeracy enable students to perform calculations, graph data, and make predictions based on this data.

Numeracy skills also help students to think critically about scientific ideas and concepts. Students can use numeracy skills to evaluate scientific claims and arguments and make evidence-based decisions. Along with this, numeracy skills acquired in science can be transferred into many real-world applications, such as in medicine, engineering, and environmental science.

Science teachers can incorporate numeracy into their activities by collecting and analysing data, conducting surveys, building models, and interpreting scientific texts. By developing numeracy skills, students can become more proficient in science and better equipped to apply scientific concepts daily.

ENGAGING WAYS TO INTRODUCE NUMERACY IN THE SCIENCE CLASSROOM

Collect and Analyse Data

Collecting and analysing data is an important part of the scientific process and requires strong numeracy skills. During first-hand investigations, students should always be encouraged to collect data and use numeracy skills to analyse and present this data.

This can be scaffolded for students depending on their skill level. Younger students can be given blank tables in which they enter their data, along with graphs with already labelled axes. As students develop their skills, this support can be removed, and students can be required to create tables and graphs from scratch and analyse their data to look for trends.

When designing investigations, one of the first steps that students should follow is clearly defining what they want to investigate and what information they want to collect. This will help them determine the purpose of the data they collect. Once this is finalised, students should choose the appropriate data collection method based on their research question. This can include surveys, experiments, observations, or interviews. Students should ensure that the data collection method is reliable and valid. This information will support the student in designing their investigation, including identifying relevant independent and dependent variables and those they need to control.

Once the method is written, students can conduct their investigation and collect the data. They should ensure that the data is collected accurately and consistently. Students can use numeracy skills to measure and record data and ensure that the data represents what is being studied. After collecting the data, students should organise and represent it. They can visually represent the data using charts, graphs, and tables. Students can use numeracy skills to make calculations, interpret data, and identify trends.

Students should analyse the data once it is organised and represented. They should identify any patterns or trends in the data and draw conclusions based on their analysis. Students can use numeracy skills to perform statistical analysis and calculate central tendency and variability measures. They should identify any patterns or trends in the data and draw conclusions based on their analysis.

The final step is to draw conclusions and communicate the findings. Students should communicate their findings clearly and concisely, using charts, graphs, and tables to support their conclusions. Students can use numeracy skills to interpret the findings and make recommendations for future research.

Conduct a Survey

Creating and conducting a survey is a great way for students to practice and demonstrate numeracy skills in science.

The first thing that students need to do in creating a survey is choose a scientific topic they want to investigate. The topic should be specific enough to generate meaningful data but broad enough to allow for diverse responses. Once the topic is selected, students then define the population and the desired size of the sample. The population is the group of people or objects the survey targets. The sample is the subset of the population that will be surveyed.

Next, students need to create survey questions that are relevant to the scientific topic. The questions should be clear, concise, and easy to understand. The students must define the type of data they want to collect, as this will determine the type of questions they write. Open-ended questions are useful for collecting qualitative data, while closed-ended questions are useful for collecting quantitative data.

Once the questions are finalised, students can start conducting the survey. Depending on the sample size and available resources, they can administer the survey in person or online. Students should ensure that the survey is conducted fairly and in an unbiased manner to ensure the accuracy of the results. After the survey is completed, students need to analyse the data. They can use numerical methods such as averages, percentages, and graphs to support their analysis and to draw conclusions. Students can also use statistical analysis to identify any patterns or correlations in the data.

The final step is interpreting the survey results and drawing conclusions about the scientific topic. Students can use the results to support or refute their hypothesis and to make evidence-based recommendations or predictions.

Building Models

Building models is an excellent way for students to practice and demonstrate numeracy skills in the science classroom.

The first step is to choose a scientific concept or phenomenon that students want to explore or demonstrate through constructing their model. This can be anything from the water cycle to the human respiratory system. After identifying the concept they will represent, students must research and plan how to build the model. Students should be encouraged to sketch a rough model design and list the materials they will need to build it so that they have a plan.

The next step is to build the model. Students will build their model using various materials; where possible, these materials should be recycled and environmentally friendly. They should follow their plan and make adjustments as necessary. When building, students may need to measure and cut materials, calculate angles and proportions or make precise adjustments to their model, allowing them to apply their numeracy skills.

Once the model is built, students should test and evaluate it. They should check that the model accurately represents the scientific concept and that it functions as intended. Students will actively use their numeracy skills to measure and record data, calculate variables such as speed and volume, and compare the model to real-world observations. If necessary, students can modify and refine the model to improve its accuracy and functionality, further demonstrating their numeracy skills to calculate and adjust the model and test it again to ensure it meets its intended purpose.

The final step is to communicate the model to others. Students can present the model to their peers or a wider audience and explain how it represents their chosen concept. They can apply their numeracy skills to create scaled diagrams and charts to support their explanations and answer questions from others.

Interpret Scientific Texts

Interpreting scientific texts is an essential skill for students in the science classroom and provides students with a way to practice and demonstrate their literacy and numeracy skills.

The first step is to identify the key terms in the scientific text. Students should identify unfamiliar terms and look up their definitions. They should also identify key concepts and scientific principles that are relevant to the text. Once the key terms are identified, students should summarise the text in their own words. They should identify the main ideas and supporting details and organise them logically. Students can also apply their numeracy skills to quantify and summarise data presented in the text.

Many scientific texts contain data and graphics such as graphs, charts, and tables. Students should analyse these to better understand the scientific concepts presented in the text. They can demonstrate their numeracy proficiency to interpret data, make calculations, and draw conclusions based on the data.

Students should connect the scientific text to real-world applications by identifying how the scientific concepts presented in the text are relevant to everyday life and how they are used in various industries and fields. Students can use numeracy skills to make calculations and projections based on the scientific principles presented in the text.

The final step is to evaluate the scientific text. Students should critically evaluate the text to identify any biases, errors, or inconsistencies. They should identify any unanswered questions and propose further research or experiments to address them. Students can use numeracy skills to evaluate the accuracy and reliability of the data presented in the text.

ENGAGING DIFFERENTIATION ACTIVITIES

Differentiation is crucial in the science classroom because it ensures that students of varied abilities and learning styles can access and engage with the material, leading to more meaningful understanding and increased academic success. As we know, the content that students are exposed to in the science classroom is often quite complex and difficult to grasp. Several activities allow students to essentially 'choose their own adventure' and, as a result, interact with the content in a way that is accessible to them.

By giving students the freedom to select tasks that suit their comfort level, educators can tailor the learning experience to each student's needs and abilities. These activities encourage self-directed learning and allow for personalised pathways to mastery, ensuring that all students can engage with science content at a level appropriate for their individual development, ultimately enhancing their overall understanding and confidence.

Tic-Tac-Toe

When students play 'tic-tac-toe', the goal is to line up three 'noughts' or 'crosses' in a line either across, down, or diagonally on the board. The tic-tac-toe differentiation method provides the opportunity for teachers to present students with a range of engaging differentiated tasks using this same concept. In this approach, a grid is created with different tasks or activities related to a specific topic or concept. Students are then given the freedom to choose which tasks they would like to complete, with the goal of completing a row, column, or diagonal of tasks, like in a game of tic-tac-toe.

This approach allows for differentiation of tasks, as each student can choose tasks that match their interests, learning style, or skill level. For example, tasks could range from creating a poster or infographic to conducting experiments or writing a research paper. By using the tic-tac-toe differentiation method, students are given a choice in their learning, promoting engagement and motivation. Furthermore, the range of tasks available allows for the development of various skills and provides opportunities for students to showcase their strengths and interests.

Create a two-minute public service announcement that addresses things that people can do to their bodies that have a negative impact on their circulatory system. Present this announcement to the class.	Make a poster that shows 5 different exercises that help to improve cardiovascular health. For each exercise, include: • the name of the exercise • an illustration of the exercise • written steps on how to perform the exercise	Research the use of medical technology for patients who are suffering from diseases of the circulatory system. Create an information brochure for patients that outlines the technology available to help them.
Research the percentage of red blood cells, white blood cells and platelets in the blood. Make a graph showing these percentages and include an explanation of the function of each component of the blood.	Draw a diagram showing how the circulatory system works. Label all of the major parts and identify their main functions in keeping the circulatory system working properly.	Research 2 diseases of the circulatory system (e.g. angina, high blood pressure or heart disease). Create a presentation explaining the causes of these diseases and how they can be prevented.
Write a paragraph explaining how the circulatory and respiratory systems are related to one another. Include a diagram in your response.	Write a short creative story that follows the journey of a red blood cell as it moves around the circulatory system, including the lungs and delivering oxygen to and collecting carbon dioxide from the organs.	Research the life and work of Doctor Victor Chang. Write a newspaper article that highlights the work he did for patients suffering from diseases of the circulatory system.

Choices - 'Think Dots'

In a **ThinkDots** activity, students must choose activities from a list that add up to a particular total score. Each activity is given a number from 1-6 (like on a dice) based on its difficulty level. Unlike the 'tic-tac-toe' activity, students do not have to choose activities in a row but simply choose a number of activities that give them the required total number of points. For example, if the student is required to earn 6 points, they may choose to do the three easiest tasks or the one hardest task. This activity allows the teacher to make individual tasks as easy or as hard as they like and make the activity as easy or challenging for students by increasing or decreasing the total score required to achieve.

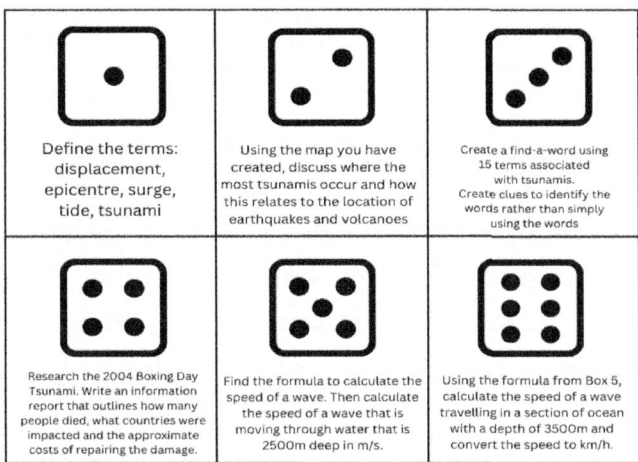

Define the terms: displacement, epicentre, surge, tide, tsunami	Using the map you have created, discuss where the most tsunamis occur and how this relates to the location of earthquakes and volcanoes	Create a find-a-word using 15 terms associated with tsunamis. Create clues to identify the words rather than simply using the words
Research the 2004 Boxing Day Tsunami. Write an information report that outlines how many people died, what countries were impacted and the approximate costs of repairing the damage.	Find the formula to calculate the speed of a wave. Then calculate the speed of a wave that is moving through water that is 2500m deep in m/s.	Using the formula from Box 5, calculate the speed of a wave travelling in a section of ocean with a depth of 3500m and convert the speed to km/h.

Traffic Light Activities

The traffic light analogy can be a useful tool for differentiating activities in the science classroom. Just like a traffic light, different activities can be categorised as 'red,' 'yellow,' or 'green' based on their level of complexity or difficulty. Red activities require a high level of prior knowledge or skill. Yellow activities include review or practice exercises that reinforce key concepts or skills. Finally, green activities include exploratory or creative projects that allow students to apply what they have learned in new and exciting ways.

The traffic light analogy is a great way to differentiate activities where students are designing their own experiments. For example, for students following the 'green light', the experiment is quite scaffolded for them, providing them with a lot of guidance and direction. In contrast, those students following the 'red light' are encouraged to be more independent in their approach to the activity and to think outside the box with minimal support.

ENGAGING COLLABORATIVE ACTIVITIES

Collaboration is paramount in the science classroom because it mirrors the essence of the scientific profession. Science is a collective effort, with scientists working together to explore, discover, and innovate. In the classroom, collaboration encourages students to pool their diverse talents, perspectives, and ideas, leading to more comprehensive problem-solving and a deeper grasp of scientific concepts.

Moreover, teamwork cultivates vital skills such as communication, critical thinking, and adaptability, which are fundamental in science and indispensable in the broader spectrum of life and work. Collaborative learning fosters an environment of support and shared knowledge and prepares students to thrive in the collaborative scientific endeavours of the real world, making it an essential component of science education.

Speed Dating

The concept of speed dating, where participants rotate quickly through brief, structured conversations, can be adapted for use in the science classroom to help students develop their communication skills and deepen their understanding of scientific concepts. In this scenario, students would be assigned a scientific concept to research beforehand and then paired with a partner for a brief discussion. Each partner would have a set amount of time to explain their concept and ask questions of the other person before rotating to a new partner.

This approach would allow students to practise explaining complex scientific ideas in a concise and understandable manner, as well as ask questions and engage in discussions to deepen their understanding of the concepts they are learning. By using this approach, students can develop important communication skills that will be useful throughout their academic and professional lives.

An example of using speed dating in a chemistry context could involve assigning each student in the class an element from the periodic table. Students research their assigned element to determine its properties, such as atomic number, valency, state and melting point. Once the students have built their individual profiles, they then 'speed date' with other elements to determine their compatibility and what kind of bond they would form. In biology, students could take on the role of organisms within an ecosystem, researching their relationships within that ecosystem. Students can then speed-date other organisms to determine what kind of relationship would exist between the two. This can lead to discussions around symbiosis and how the relationships between organisms are vital for ecosystems to survive.

Jigsaw Activity

The jigsaw classroom technique is an effective way to promote communication and collaboration among students in the science classroom. In this approach, students are divided into small groups, and each is assigned a different component of a larger concept or topic. After each group has researched and learned about their assigned component, students are reorganised into new groups containing one member from each initial group. In these new groups, students then share what they have learned about their assigned component, allowing for a comprehensive discussion of the larger concept or topic.

Students can develop their communication skills by presenting and explaining their research findings to their peers by participating in this jigsaw activity. Furthermore, the discussion during the activity promotes critical thinking and allows students to strengthen their understanding of scientific concepts by building upon the knowledge of others.

The jigsaw technique can be used whenever your students are exploring a wide range of similar concepts within a topic and you are short on time. One example of a jigsaw activity is a research task on medical imaging technology where students are divided into eight 'home' groups and are allocated a medical imaging technology to research. Individuals within the 'home' group become experts in their particular technology based on the requirements outlined in the task. Once each group has completed their research, students are regrouped into four groups of eight.

Each expert shares their information to help the others summarise the different medical imaging technologies available. Another example in a chemistry context could be assigning each home group a historical model of the atom that they will research together and become experts on. After researching, these students will share their findings with another student from the other home groups to build a collaborative understanding of all atomic models throughout time.

Think-Pair Share

The Think-Pair-Share strategy is a collaborative learning technique that encourages active participation and engagement among students. This method involves three stages: first, the students are asked to think individually about a particular topic or question the teacher presents. Then, they pair up with another student to discuss their thoughts and ideas. Finally, the pairs share their ideas with the whole class. This strategy promotes critical thinking, problem-solving, and communication skills and allows students to learn from each other. It also helps students develop their confidence and ability to articulate their thoughts and ideas in a group setting.

The Think-Pair-Share strategy can be used in various topics; however, one way to use this strategy in your classroom includes having the students create a concept map of human body systems. To start, students will work individually for a short time, writing down what they already know about each body system in relation to the structures and their overall functions. After this time, students will share their mind map with the person next to them and add any missing information to their original mind map. Lastly, in small groups, students will go through their mind maps to ensure they all have the same information. Teachers can observe the level of detail produced in the student's mind maps to understand their prior knowledge of the body systems.

An example of the Think-Pair-Share strategy in physics could be used in the introduction of a unit on waves. To start, students can be asked to think of the different forms of waves they know and how these waves are used. Students can then share their initial thoughts with their partner before joining with another pair of students to share their thoughts to collaboratively identify their understanding of waves.

Design a Board Game

Collaboratively creating a board game can be a fun and engaging way for students to learn and understand scientific concepts. When students design a board game, they must research and understand the scientific concept they are trying to represent. This helps reinforce their understanding of the topic and increases their knowledge retention. Board games can engage multiple learning styles, including visual, auditory, and kinesthetic learners. By designing and playing a board game, students can learn through various mediums, such as reading, listening, and hands-on activities.

To design a board game, students must think critically to ensure the game is both entertaining and educational while incorporating scientific concepts. This helps to develop their creativity and problem-solving skills, which are valuable in science and other subjects. Working together to design and test their games helps to develop students' communication and collaboration skills, which are important in science and other areas of life.

Board games are often enjoyable and provide a memorable learning experience. By creating a board game, students are more likely to remember the scientific concepts they learned and may even develop a passion for the subject.

Creating a board game to learn about the different stages of the water cycle and the processes involved is one example of a board game that students can create in science. With this particular activity, students should be given as much freedom as possible to create their board game while being encouraged to be creative and add fun elements to the game, such as trivia questions or challenges that players must complete.

FORMATIVE ASSESSMENT

Formative assessment is the type of assessment that is designed to help teachers understand how their students are progressing and to identify areas where they may need additional support. It is an ongoing process that occurs throughout instruction and is used to inform and adjust teaching and learning strategies. In the science classroom, formative assessment can be an invaluable tool for improving student learning and achievement.

One of the main benefits of formative assessment in the science classroom is that it allows teachers to identify and address misunderstandings or gaps in student knowledge early on. By regularly checking in on student progress and gathering student feedback, teachers can determine which concepts students struggle with and provide targeted support to help them master the material. This is particularly important in science, where a thorough understanding of foundational concepts is necessary for building more complex understanding in subsequent units or courses.

Formative assessment can also help to engage and motivate students. By involving students in the assessment process and providing opportunities for them to self-assess and set learning goals, teachers can encourage students to take ownership of their learning and become more invested in the material. This can lead to increased student engagement and a more positive classroom environment.

Another benefit of formative assessment in the science classroom is that it can help foster critical thinking skills. Teachers can encourage students to develop the skills necessary to evaluate and analyse scientific evidence by asking students to explain their thinking and providing opportunities to engage in tasks such as those often found in inquiry-based learning. This is important for helping students to become more independent learners and to develop a deeper understanding of scientific concepts.

Formative assessment can also be a useful tool for differentiating instruction. By gathering ongoing feedback on student progress and adjusting teaching strategies based on that feedback, teachers can better meet the needs of individual students and ensure that all students can succeed.

Using Technology for Formative Assessment

Kahoot (kahoot.com) is a game-based learning platform that allows educators to create and share interactive quizzes, surveys, and discussions with their students. It is a popular tool in classrooms and educational settings worldwide, as it engages students in learning while providing a fun and interactive experience. Kahoot can be used in various subjects and is flexible enough for use in various settings, including in-person and online classrooms, workshops, and professional development sessions.

To use Kahoot, educators are able to create their own quiz or survey or use one that has already been created by others. Students then access the quiz using a unique PIN code on their own devices. As the quiz progresses, students answer questions and see their results in real-time. Kahoot is a powerful tool for promoting interactive and engaging learning in classrooms and educational settings.

Plickers (plickers.com) is a student response system that allows teachers to assess student understanding in real-time using physical cards and a mobile app. Each student is assigned a unique card with four multiple-choice answer options represented by different shapes. To respond to a question, students hold up their cards with the correct shape facing the teacher. The teacher uses the Plickers app to scan the room and quickly collect and record the responses from all students.

Plickers is a useful tool for teachers because it allows them to gauge student understanding and adjust their lessons quickly. It also encourages student participation and can help to identify misunderstandings or areas where students may need additional support. Plickers can be used in various settings and is a simple and convenient way for teachers to assess student learning and provide targeted support to individuals or small groups.

Quizlet (quizlet.com) is a popular study tool that allows users to create and share flashcards and other study materials. It offers a variety of features, such as the ability to create and share study sets, practice quizzes, and interactive games to help students learn and retain information. Quizlet also has a mobile app, which makes it convenient for students to study on the go.

One of the key features of Quizlet is that it allows users to collaborate, making it a useful tool for group study sessions. In addition to its study tools, Quizlet has an extensive library of pre-made study sets created by other users, which can be a helpful resource for students studying for a specific subject or exam. Overall, Quizlet is great for students looking to improve their study habits and retention of information.

Google Forms is another valuable tool for teachers to use for formative assessment in the classroom. One way that teachers can use Google Forms is by creating quizzes or surveys to assess student understanding of a particular concept or topic. Teachers can include a variety of question types, such as multiple-choice, short-answer, and essay questions, to get a comprehensive understanding of student knowledge and skills. Google Forms provides the functionality of automatically marking a range of question styles and creates a Google Sheet of the student's responses, which can be used to analyse the results easily.

Blooket (blooket.com) is a versatile digital tool with immense potential for formative assessment in the science classroom. This platform allows teachers to create interactive, multimedia-rich content that aligns perfectly with science education's dynamic and diverse nature. With Blooket, educators can craft engaging quizzes, surveys, and interactive lessons that provide instant feedback to gauge student comprehension and adapt instruction accordingly. Its flexibility enables the integration of images, videos, and simulations, fostering a deeper understanding of complex scientific concepts.

Blooket makes assessment more interactive and engaging and provides educators with valuable insights into their student's progress, allowing for targeted, data-driven adjustments to content and teaching strategies. In the ever-evolving landscape of science education, Blooket emerges as a powerful tool for enhancing formative assessment practices and promoting a more effective and engaging learning experience.

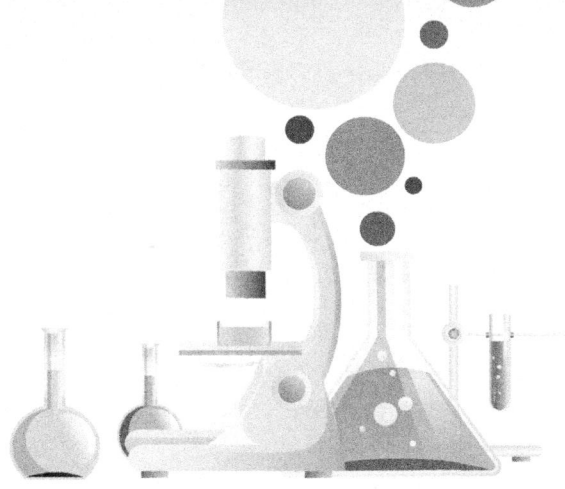

CREATING ENGAGING OPPORTUNITIES WITH TECHNOLOGY

The use of technology in the science classroom has become increasingly common in recent years. Technology can be used to enhance traditional teaching methods, provide students with access to a range of multimedia resources, and promote engagement and active learning.

TAKING ADVANTAGE OF THE SAMR MODEL IN SCIENCE

Technology allows science educators to reimagine how they present information to their students and how their students create artefacts to demonstrate their learning. The **SAMR model** is a framework for evaluating and enhancing the use of technology in education. At its core, the model is centred around the idea that technology can be used to augment, modify, or redefine traditional teaching and learning activities. The letters SAMR stand for 'substitution', 'augmentation', 'modification' and 'redefinition'. Dr Ruben Puentedura developed this method to help teachers and educators understand how to meaningfully integrate technology into their classrooms.

The substitution level involves using technology to simply replace an existing task or activity without necessarily improving upon it. In contrast, augmentation involves using technology to enhance or add value to an existing task or activity. An example of substitution would involve using a computer to type out an assessment rather than writing it by hand, while augmentation could involve using a calculator to solve numerical problems, as it makes the task easier and more efficient but does not fundamentally change what is completed.

The modification level involves using technology to significantly alter or transform an existing task or activity. For example, using a virtual reality headset to take students on a virtual field trip would be considered modification, as it significantly changes the task design.

Finally, the redefinition level involves using technology to create new, previously unimaginable tasks or activities. For example, using 3D printing to create physical objects from digital designs would be considered redefinition, as it allows for the creation of completely new and innovative products.

The value of the SAMR model lies in its ability to help educators understand how to effectively integrate technology into their classrooms in a way that enhances and improves the learning experience for their students. By evaluating the tasks and activities they use technology for, educators can determine which level of the SAMR model they are operating at and consider ways to move up to higher levels, ultimately using technology to create more meaningful and transformative learning experiences for their students.

AR in the Science Classroom

Augmented reality (AR) is a technology that allows users to view and interact with digital content in the real world. It can potentially revolutionise how science is taught in the classroom by providing students with immersive and interactive learning experiences.

A range of apps exist that provide access to virtual reality experiences through a smartphone or tablet. These can be accessed through various app stores and downloaded to students' devices. Many of these may require a small payment to access; however, the quality of some of these apps is quite incredible, and it may be possible to claim the purchase back on tax as a teaching aid (just check with your tax accountant first!).

The **SkyView app** is one of my favourite apps for viewing the night sky. This app can be purchased for a small fee; however, the free version also does a great job. By simply pointing this app at the sky, students are able to identify the location of all celestial bodies while also being provided with interesting information about them. It can also be used to identify bodies that are not visible in the night sky due to being beyond the horizon.

In addition to AR apps and games, teachers can also use AR to enhance traditional teaching methods. For example, a teacher might use AR to display 3D models of scientific concepts, such as atoms or cells, allowing students to better understand and visualise these complex structures. This can be especially useful for students who struggle with traditional teaching methods or have English as an additional language, as the visual elements provide these students with a more interactive and engaging way to learn.

AR can also be used to bring science to life in the classroom by allowing students to experience scientific concepts in a more hands-on and immersive way. For example, an AR field trip could allow students to visit a virtual ecosystem and learn about the different plants and animals that live there, or an AR lab could allow students to conduct virtual experiments and explore scientific concepts in a safe and controlled environment.

Overall, the use of AR in the science classroom has the potential to enhance students' learning experience greatly. It allows them to learn in a more interactive and immersive way and can help bring scientific concepts to life in a way that traditional teaching methods cannot.

VR in the Science Classroom

Virtual reality (VR) in science education can also significantly transform the learning process, offering students immersive, interactive experiences that promote a deeper grasp of scientific concepts and ignite a heightened sense of curiosity. One of the main benefits of using VR in the science classroom is that it allows students to engage with concepts in a much more interactive and engaging way than traditional methods such as lectures or textbook reading. VR allows students to experience scientific phenomena firsthand, giving them a deeper understanding of the material.

For example, VR can simulate a trip to the moon or the bottom of the ocean, allowing students to explore these environments and learn about the scientific concepts related to them. VR can also be used to simulate complex chemical reactions or biological processes, helping students understand these concepts in a more visual and interactive way.

One extremely easy way to incorporate VR into the science classroom is through the use of YouTube 360° videos and a headset. These videos are freely available, and cardboard headsets can be purchased online for a reasonable price. A great way to introduce a physics unit on motion and acceleration is to have a student watch a 360° video of a rollercoaster while sitting on a rolling chair - ideally, with the video being cast onto a projector for the rest of the class to see simultaneously. While the student is experiencing the sensations through the headset, move the chair around to give the student a truly immersive experience.

Additionally, VR can be used to create immersive learning environments that allow students to collaborate on scientific projects. This can foster teamwork and problem-solving skills and encourage critical and creative thinking.

However, it is important to note that VR is not a replacement for traditional teaching methods, and it should be used in conjunction with other methods, such as lectures and hands-on experiments. Additionally, it is important to ensure that students have access to the necessary VR technology and are trained on how to use it safely.

Overall, VR technology has the potential to significantly enhance the way science is taught in the classroom. By providing immersive, interactive learning experiences, VR can help students to better understand and engage with complex scientific concepts, fostering a deeper appreciation for science and a stronger foundation for future learning.

Virtual Simulations

Virtual simulations allow students to explore scientific concepts in a safe, controlled environment, making it possible to conduct experiments and explore phenomena that would be difficult or dangerous to replicate in a traditional classroom setting.

PhET (phet.colorado.edu) is one of the most well-known online simulation tools available to science educators worldwide. Designed and maintained by the University of Colorado, PhET provides a range of simulations for a number of science topics across all strands. These activities are free to use with your students and can be embedded into external resources such as websites and blogs.

Other virtual simulation sites exist that will allow students to explore different concepts and conduct a range of scientific investigations from the comfort of their laptops. Doing a Google search for terms such as 'online or virtual science investigations' will help you find these. Many free ones are available for use in an educational setting, along with paid experiences.

FLIPPING THE CLASSROOM

Flipped learning is a teaching method in which students engage with the course content by watching online lectures or videos at home and then participating in discussions and activities in class. This student-centred approach to learning has gained popularity in recent years, particularly in science classrooms, due to its potential to enhance student learning and engagement.

Implementing a flipped classroom model allows teachers to shift the lower end of Bloom's taxonomy out of the classroom, allowing them to be present with their students as they face more difficult activities, apply knowledge and delve more deeply into topics. By completing a range of easily achievable activities at home, students interact with Bloom's 'remember and understand' levels outside the classroom without their teacher, reserving time spent in class for the higher-order levels of thinking, including creating, evaluating, analysing, and applying. These activities usually take longer to complete and often require the support and input of the classroom teacher.

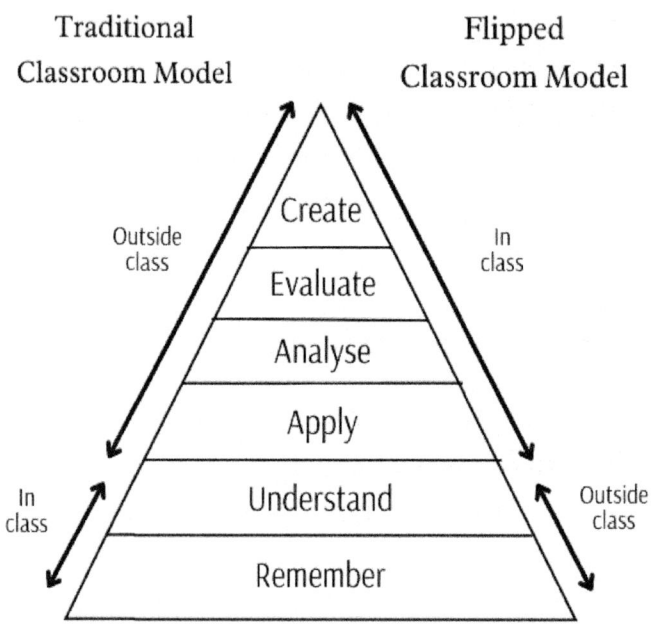

The 'traditional' classroom model sees students often take notes from the board or read the information before completing project tasks at home. Flipping this process allows for those higher-order activities to be explored where students feel safe and supported. Subjects that consist of educational content that falls within these lower levels of Bloom's taxonomy may benefit the most from a shift toward the flipped classroom model of teaching.

YouTube changed how the world used the Internet when Jawed Karim uploaded the first YouTube video on April 23, 2005. Combined with the availability of low-cost computer-based video capture capabilities, we have seen dramatic transformations in how educators and students experience education. Team these YouTube videos with many other technological tools and learning management systems, and educators are creating an open and social environment for their students to interact in and with. These types of tools have shifted the web towards an interactive, two-dimensional environment where participation is the key.

A large range of tools exist that allow educators to create, produce, upload, and distribute video content themselves. These tools provide an opportunity to teach in ways that were not possible in the past and have been transformative over the last decade. Some of these tools include screencasting software such as Camtasia and Screencast-o-matic, both of which allow educators to create a video of the activity that is taking place on their computer screen. By using these screencasting and video production tools, teachers can produce video content of lectures that students can consume in their own time, outside of the classroom environment.

Flipping the science classroom can be daunting; however, there are some tips for when you take the leap:

- Have clearly defined goals and expectations: introducing the flipped classroom requires buy-in from all parties involved. Students need to understand that they will be taking responsibility for a large portion of their learning, so it is beneficial to ensure that some guidelines are put in place before diving in. It is also a great idea to have the support of the student's parents when embarking on a flipped learning journey. Providing them with insights into how your classroom will run and the model's potential benefits will help them navigate their child's concerns when taking on greater responsibility for their learning.

- Create clear and concise at-home tasks: because students will be engaging with new content at home, you need to ensure that the delivery of this information is packaged in a way that students can understand. If you are going to be creating videos, keep them short. The ideal time frame is 1-2 minutes multiplied by the student's year level (e.g. if a student is in Year 7, a video should go for no longer than 7-10 minutes). Ideally, if these can be made shorter, the better. Providing opportunities to engage with the video content in class with guidance at the beginning of the flipped learning journey can help set the students up for success by modelling how you wish them to interact with the content.

- Ensure that the face-to-face time is spent on engaging activities: it is important that you do not re-visit all of the content from the at-home tasks in class. This will discourage students from engaging with this content before they arrive, as they can simply get the information in class. The face-to-face time should be used to dive deeper into the concepts, requiring students to apply their knowledge in a range of ways. Group tasks, hands-on activities, and project work are ways of engaging face-to-face class time.

- Don't flip the whole course: you must identify the content you can move out of the face-to-face space and what you will still need to address with your students. This comes down to knowing your subject content and the nuances that must be approached with each particular part of a topic.

Online Curriculum-Aligned Resources: Education Perfect

When embarking on a flipped learning journey, creating all resources from scratch is unnecessary. Teachers can lean on a wide range of tools to support their flipped classroom, such as those in the Education Perfect (EP) platform. EP is an online learning platform designed to help students improve their skills and knowledge in various academic subjects. The platform offers curriculum-aligned resources for a number of curriculum boards. This content is presented using scaffolded lessons, practice exercises, quizzes, games, and videos to help students learn in a fun and engaging way. It also provides teachers with tools to track students' progress and identify areas for improvement.

One of the key features of EP is its adaptive learning technology, which tailors the content and difficulty of the exercises to the individual student's needs. This means that students can work at their own pace and receive personalised feedback on their progress. The platform also offers a range of reporting tools, allowing teachers to monitor students' progress and identify areas of weakness.

One of the key benefits of using EP in a flipped classroom is that it can help promote active learning and student engagement. By providing students with access to a range of multimedia resources, EP can help to make scientific concepts more accessible and engaging, particularly for students who may struggle with traditional classroom methods.

Additionally, EP can be used to support differentiated instruction, providing students with personalised learning experiences based on their individual needs and interests. Teachers can use EP to create custom learning paths tailored to each student's strengths, weaknesses, and learning styles, providing them with the support and challenge they need to achieve their learning goals.

Finally, EP provides opportunities for ongoing assessment and feedback. By tracking student progress and providing real-time feedback on their performance, teachers can use EP to identify areas where students may need additional support or challenge and to adjust their instruction accordingly.

COLLABORATION WITH TECHNOLOGY

Technology can also be used to promote collaboration and communication in the science classroom. For example, online discussion boards or collaborative document editing tools can be used to encourage students to work together on projects, share ideas and provide feedback. These tools can help students build knowledge and promote critical thinking skills, as students can discuss and evaluate scientific concepts with their peers.

Padlet (padlet.com) is an amazing tool for collaboration in the classroom. With this tool, teachers can create an online sticky note wall that allows students to post their comments, questions, and resources in one place that is easily accessible to everyone. When a new Padlet is created, it can be kept private, made public or shared with a specific group of people, allowing teachers to ensure the digital safety of their students.

Padlet allows for sharing almost any file type, including text, images, videos and links, that board viewers can comment on and 'like'. By creating a shared space where students can share their ideas and resources, Padlet can help promote peer-to-peer learning and support the development of communication skills.

Teachers can also harness the power of Padlet to generate self-directed learning opportunities, similar to hyperdocs (discussed below). By creating a Padlet board related to a specific topic or concept, teachers can provide students with a range of resources and activities that they can explore at their own pace. This approach can be particularly effective for students who need additional support to master scientific concepts, and it supports different learning styles.

Hyperdocs are digital documents that contain a range of multimedia resources and activities related to a specific topic or concept. The concept of Hyperdocs was introduced in 2013 by educators Lisa Highfill, Kelly Hilton and Sarah Landis. Hyperdocs can provide students with a range of engaging, differentiated tasks that promote active learning and collaboration. They can be designed to incorporate a range of multimedia resources, such as videos, podcasts, and interactive simulations, to help students explore scientific concepts in depth.

Hyperdocs can be used in a variety of ways in the science classroom. For example, teachers can use Hyperdocs to provide students with self-paced learning opportunities, allowing students to work through the material at their own pace and explore topics in greater depth. Additionally, Hyperdocs can provide students with differentiated learning opportunities, allowing them to choose tasks that match their interests, learning styles, or skill levels. This approach promotes student engagement and motivation, as they are given a choice in their learning.

Hyperdocs can also be used to promote collaboration among students. By incorporating collaborative tasks, such as online discussions, group projects, or peer review activities, Hyperdocs can help students develop their communication and collaboration skills.

Furthermore, by providing students with a range of multimedia resources, Hyperdocs can help promote digital literacy skills and prepare students for the demands of the 21st-century workplace.

Lastly, Hyperdocs can support student-centred learning and promote inquiry-based learning in the science classroom. By incorporating open-ended questions and opportunities for students to explore scientific concepts in depth, Hyperdocs can help students develop their critical thinking and problem-solving skills. Hyperdocs are a versatile and effective tool that can be used in various ways to promote active learning, collaboration, and inquiry in the science classroom.

MORE ENGAGING TECH TOOLS

ThingLink

ThingLink (thinglink.com) is a web-based platform that allows users to create interactive images and videos. With ThingLink, users can easily add interactive elements such as text, links, and multimedia to their images and videos, creating a more engaging and interactive experience for their audience.

One of the key benefits of educators using ThingLink in the science classroom is that it can help to make complex scientific concepts more accessible and engaging for students. Providing students with access to interactive resources incorporating multimedia elements, ThingLink can help promote active learning and critical thinking, allowing students to explore scientific concepts more effectively and engagingly.

Students can also create their own ThingLinks to demonstrate their understanding of a concept while engaging in active learning and critical thinking. By selecting relevant multimedia resources and embedding them into an interactive image or video, students must think carefully about the scientific concepts they are trying to convey and how best to present them. This process can deepen their understanding of scientific concepts and encourage them to think more creatively about communicating their ideas to others.

Allowing students to create their own ThingLinks can give them a sense of ownership and autonomy over their learning. By allowing students to create their own interactive resources, teachers can provide them with a sense of agency and control over their learning process, which can be motivating and empowering. This can foster a more student-centred and personalised learning environment. By sharing their ThingLinks with others, students can develop their communication and collaboration skills. By presenting their interactive resources to their peers, students must learn how to communicate complex scientific concepts in a clear and engaging way, and they must work collaboratively to receive and incorporate feedback from others. ThingLink could also be used as an assessment tool for students to showcase their understanding of scientific concepts creatively.

An example of how ThingLink can form part of a learning experience for students involves researching an animal's adaptations and creating a ThingLink to teach other people about them. Students could research at least two behavioural, two structural and two physiological adaptations and describe these using ThingLink. Each adaptation explanation could require a multimodal element, such as a photograph or video and an explanation of how it helps the organism survive in its natural habitat, thereby encouraging them to include text, images, and videos in their creations. ThingLinks can then be embedded into other tools, such as websites, to allow students to demonstrate their understanding further.

Mentimeter

Mentimeter (mentimeter.com) is a digital tool that allows teachers to create interactive presentations and engage their students in real-time. It offers a variety of question formats, including multiple-choice, open-ended, and scales. Students can respond to the questions through their mobile devices or computers, and the results are instantly displayed on the presentation screen.

One way to use Mentimeter in the science classroom is to create pre-assessment quizzes to gauge students' understanding of scientific concepts before starting a new unit. Teachers can use multiple-choice questions to test their students' knowledge and understanding of the topic. This diagnostic assessment can help the teacher tailor their lessons to meet their student's needs and help students identify areas where they need more support.

Another way to use Mentimeter in the science classroom is to create interactive activities that encourage students to think critically and creatively about scientific concepts.

For example, teachers can use open-ended questions to prompt students to think about the potential applications of a scientific concept or theory. Students can then share their ideas with the class, sparking discussions and debates.

Mentimeter can also be used to gather student feedback on their learning experience. Teachers can use scale questions to ask students about their confidence in their understanding of a particular topic or their engagement with a particular lesson. This can help teachers adjust their teaching approach to better meet their students' needs.

Finally, Mentimeter can be used as a formative assessment tool to track student's progress throughout a unit or course. Teachers can use multiple-choice questions to test students' knowledge and understanding at different stages of the learning process and then use the results to adjust their instruction as needed. This can help ensure that students master the content and achieve their learning objectives.

QR Codes

QR codes can be used in the science classroom as an interactive and engaging way to provide students with additional information, resources, and activities. Teachers can create QR codes that link to various multimedia resources, including videos, articles, infographics, and simulations. By scanning the QR codes with a smartphone or tablet, students can access these resources quickly and easily without typing in URLs or searching for information online.

QR codes can also be used to create scavenger hunts or interactive learning activities. Teachers can create a series of QR codes that lead students on a scavenger hunt around the classroom or school, encouraging them to explore different scientific concepts and discover new information along the way. QR codes can also link to interactive quizzes or assessments, allowing students to test their understanding of scientific concepts and receive instant feedback on their performance.

QR codes can be pathways to activities that promote active learning and critical thinking. By providing students with access to a range of multimedia resources, QR codes can encourage students to explore scientific concepts in a more interactive and engaging way. This can help deepen students' understanding of complex scientific ideas.

Memes

Memes, which are images or videos modified with humorous captions or annotations, can be used in the science classroom to enhance student engagement and promote deeper learning. By using memes relevant to scientific concepts, teachers can create a more relaxed and informal learning environment that can help reduce student anxiety and promote a sense of fun and enjoyment.

Using memes in the science classroom can help to make complex scientific concepts more accessible and understandable to students. Using humorous captions or annotations that relate to the scientific content, memes help simplify complex concepts and make them more memorable and relatable for students. This can be particularly effective for visual learners, who may find it easier to remember information that is visually engaging.

By encouraging students to create their own science-related memes, teachers can provide them with a fun and engaging way to express their understanding of scientific concepts and demonstrate their creativity and critical thinking skills. Creating memes can be a fun way for students to develop revision material for concepts that may otherwise be 'dry' and not overly engaging for students.

As a parting gift for one of my Year 12 Biology classes, I created a slide deck that had over 100 memes that related to almost all of the dot points from the syllabus. The students loved it as it reminded them of our time together, but it also was a valuable tool to help them revise for their exams.

Digital Escape Rooms

Digital Escape Rooms are an innovative and exciting way to engage students and promote their learning in the science classroom. Using digital tools and software, teachers can create interactive escape rooms that challenge students to use their critical thinking and problem-solving skills to solve a series of puzzles or challenges related to scientific concepts.

One of the key benefits of using digital escape rooms in the science classroom is that they provide a fun and engaging way to learn and reinforce scientific concepts. Students work together to solve puzzles, complete challenges, and unlock clues while learning about scientific concepts and phenomena. The gamified nature of digital escape rooms can motivate students and make learning more enjoyable.

Digital escape rooms also encourage collaboration and teamwork among students as they work together to solve puzzles and complete challenges, promoting important social and communication skills and enhancing students' understanding of scientific concepts through peer-to-peer learning.

Another advantage of using digital escape rooms in the science classroom is that they allow for differentiation and personalisation of learning. Teachers can create escape rooms tailored to different students' needs and interests, allowing them to work at their own pace and level of difficulty.

There are several tools that teachers can use to create digital escape rooms in science. Some popular options include:

- **Google Forms** - Teachers can use the 'quiz' feature of Google Forms to create a series of questions or challenges related to scientific concepts. They can then use the 'response validation' feature to create unlocking clues or hints when students answer questions correctly.

- **Basic website builder** - When creating a website, creating the back end of the URL is often possible. Using this, teachers can create a digital escape room where students find clues by carrying out various activities and use the code they work out to input into the URL to access the next challenge.

- **Breakout EDU** (breakoutedu.com) - This digital platform provides a library of pre-made escape room challenges for a range of subjects, including science. Teachers can customise these challenges to fit the needs of their students and track their progress and completion. Breakout EDU is a paid resource.

- **EdPuzzle** (edpuzzle.com) - Teachers can use EdPuzzle to create interactive video lessons that include questions and challenges related to scientific concepts. Students must answer these questions correctly to unlock clues or hints that help them progress through the escape room. EdPuzzle can be integrated with a number of popular learning management systems.

- **ThingLink** (thinglink.com) - This platform allows teachers to create interactive images or infographics that include links to questions, challenges, or clues related to scientific concepts. Students can click on these links to progress through the escape room.

AI Tools - Canva And DALL·E 3

Artificial intelligence (AI) is transforming education in many ways. One of the most significant impacts is AI's ability to personalise each student's learning experience. By analysing student performance and behaviour data,

AI-powered education platforms can provide tailored recommendations and interventions to help students learn at their own pace and in ways that work best for them. This personalised approach can lead to better learning outcomes and increased engagement and motivation.

Students can creatively use AI in science by incorporating it into their experiments and projects. For example, students can use AI to analyse large amounts of data collected during an experiment and to identify patterns or trends that may not be apparent to the naked eye. They can also use AI to create models and simulations of complex scientific phenomena, such as climate change or the spread of diseases.

Canva (canva.com) - has released a series of products that allows primary and secondary teachers and students to harness the power of AI to create images, videos, presentations and text for free. **Dall-E3** (openai.com/dall-e-3) is another AI-powered tool that students can use to create unique and visually stunning images related to scientific concepts.

Both of these tools require students to enter a text-based prompt. Students can input a description of a scientific concept or idea, and the tool will generate an image representing it. This can be a powerful tool for helping students visualise complex scientific concepts, such as molecular structures or geological formations.

In addition to helping students visualise scientific concepts, AI tools can also be used as a creative tool for science projects and presentations. Students can use these tools to generate images representing their scientific findings or data sets, making their presentations more engaging and visually appealing.

Both Canva and Dall-E3 can also be used to explore the intersections between art and science. Using these tools to create artistic representations of scientific concepts or phenomena can help students develop a deeper appreciation for the beauty and complexity of science.

Social Media Post Generators

While it is not appropriate to create fake social media posts from famous people, there are ways to use social media post generators to engage students in science by creating hypothetical posts that align with the famous person's interests and accomplishments. For example, you could create a post from Albert Einstein that discusses his theory of relativity and how it has revolutionised our understanding of the universe.

To engage students in this activity, ask them to research famous scientists and explore background information about them and their contributions to science. Then, you could have them work either individually or in groups to create hypothetical social media posts that align with the famous person's interests and accomplishments. The posts could include text, images, and videos highlighting key scientific concepts and discoveries related to the person's work. This approach encourages students to delve into the intricate details of scientific breakthroughs, fostering a deeper understanding of the challenges, triumphs, and societal impact associated with various human endeavours.

To take this activity a step further, you could have students present their posts to the class and explain the scientific concepts and discoveries they highlighted. This can help students develop their communication skills and deepen their understanding of key scientific concepts. Additionally, it can spark discussions and debates among students about the role of science in society and the importance of scientific discovery.

Several websites provide pretty authentic-looking social media post generators; however, one where you can find templates for a range of social media platforms is **Zeoob** (zeoob.com). Zeoob provides templates for Facebook, Instagram and Twitter/X posts and allows the students to create conversations between two or more individuals via the WhatsApp template. These conversational posts could be used to represent debates between two conflicting parties in historical events related to science. As with all online tools, it is important to supervise students using Zeoob as it is an open-access tool.

Video Content

Videos can visually represent scientific concepts that may be difficult to explain with just words or images and can help students better understand complex ideas and theories. Videos can be a highly engaging way to capture students' attention and encourage them to participate in learning. Students may also be more likely to retain information presented in video format than traditional lectures or textbooks.

With platforms like **YouTube** and **ClickView** (clickview. com.au), teachers have access to a vast library of free or low-cost educational videos to supplement classroom instruction. This makes video an affordable tool for teachers who may not have access to other resources or technology. These resources can be accessed anytime and anywhere, making them a convenient tool for teachers and students and ensuring that students who miss a class or need to review a concept can easily access the information they need.

Videos can also provide students with various perspectives and viewpoints on scientific topics. This can help students develop a more well-rounded understanding of the subject matter and encourage critical thinking and debate.

There are also several benefits to having students create their own videos to demonstrate their understanding of science concepts. First and foremost, creating a video requires students to engage with the material in a more active and hands-on way. They must think critically about the concepts they are trying to convey and find creative ways to present the information in a way that is clear and easy to understand. This process helps solidify their understanding of the material and allows them to identify gaps in their knowledge. Additionally, creating a video can be a fun and engaging way to learn, and it can help foster a sense of ownership and pride in their work.

Another benefit of having students create videos is that it can help to develop their communication and collaboration skills. When creating a video, students must work together to plan and organise their ideas, assign roles and responsibilities, and ultimately produce a final product. This process requires them to communicate effectively with one another and work together towards a common goal. Furthermore, creating a video allows students to share their work with a larger audience, which can help to build their confidence and give them a sense of accomplishment.

Stop-Motion Animation

Stop-motion animation is an innovative and engaging way to promote student learning and understanding of scientific concepts in the classroom. By using stop-motion animation techniques, students can create videos that demonstrate their understanding of scientific phenomena, experiments, and concepts in a fun and creative way.

Creating stop-motion animations in the science classroom allows students to visualise complex scientific concepts in a tangible way. Students can better understand the cause-and-effect relationships underpinning scientific concepts by creating short animations that depict the different steps of an experiment or the sequence of a scientific process. Stop-motion animation also encourages collaboration and teamwork among students as they work together to plan, storyboard, and create their videos. This can help promote important social and communication skills and enhance their understanding of scientific concepts through peer-to-peer learning.

Meiosis and mitosis, life cycles, DNA replication, phases of the moon, the act of forces on an object and chemical reactions are all concepts that make great stimuli for students to create stop-motion animations to share their learning.

Make a Hologram

Holograms use a photographic technique that scatters the light from an object and projects it in a way that appears three-dimensional. They appear in movies such as 'Star Wars' and 'Iron Man', but the technology has not quite caught up to movie magic yet. Your students, however, can make a hologram projector using basic items found around the home or school to create amazing three-dimensional videos using their phone or tablet.

Students require a thick piece of plastic - suggested sources of this include a CD case or thick overhead transparency/binder cover and a pair of strong scissors or a Stanley knife. The CD case can be quite difficult to cut and can become quite sharp when doing so, so depending on your students and the materials you have available, safety steps must be followed in creating these projectors.

The first step is to draw a trapezium on the plastic using the dimensions 1 cm by 3.5 cm by 6 cm. Graph paper can help the students create their trapezium template and they can transfer it onto the plastic using a marker. The students will need to cut out four of these trapeziums and use clear tape to stick them together into a pyramid shape.

To create the hologram, find an appropriate video using a service such as YouTube and place the projector in the middle of the screen. Turning the lights off will give the best results, allowing the light to be scattered in the necessary way to create the three-dimensional hologram.

Students can also use technology to create their own videos that can be used with the hologram projector. Using a phone or camera to film their video, students can then import this into PowerPoint (or similar) four times and arrange each video version in a fashion that produces an 'X' with the base of the videos towards the middle of the screen. Once saved, the student can use their hologram projector to create a unique three-dimensional hologram of themselves as a digital artefact that showcases the student's understanding of a concept.

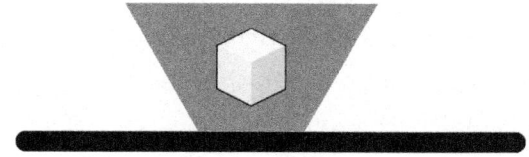

Flip

Flip (flip.com) is a video discussion platform that allows teachers to create online discussion forums and engage their students in video-based conversations. Students can record and share videos of themselves discussing a particular topic or answering a question posed by the teacher and can also watch and respond to their classmates' videos. Flip can be useful for promoting student engagement and facilitating meaningful class discussions.

One way to use Flip in the science classroom is to assign students a topic or question related to a particular concept and ask them to record and share a short video in response. For example, a teacher might ask students to discuss the potential effects of climate change on local ecosystems or explain the mitosis process in their own words. By watching and responding to each other's videos, students can engage in thoughtful discussions and learn from one another's perspectives.

Another way to use Flip is to create video-based labs or experiments that can be conducted at home. For example, a teacher might ask students to use common household materials to conduct an experiment related to the properties of water and to record and share a video of their findings. This can be a fun and engaging way to help students learn scientific concepts, develop their analytical skills and share knowledge collaboratively.

Finally, Flip can be used to facilitate virtual group projects or presentations. Students can record and share videos of their presentations or group discussions, which can be viewed by other class members or shared with outside audiences. This can be a useful tool for classes that are engaging in collaborative projects or research.

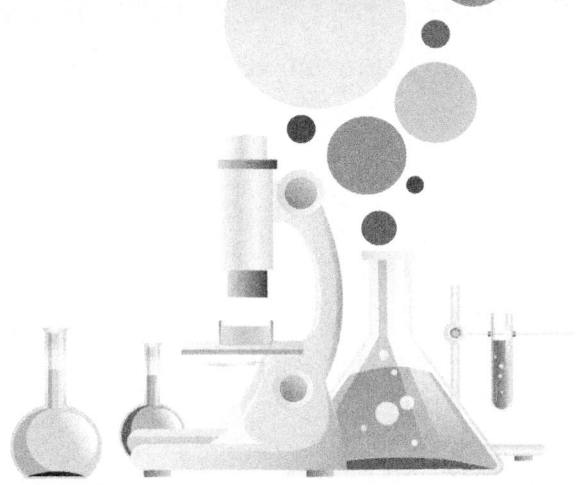

CREATING ENGAGING HANDS-ON LEARNING EXPERIENCES

CHEMISTRY ACTIVITIES

Ice Cream in a Bag

Context: In this experiment, students will investigate the science behind making ice cream using a simple and fun method known as 'ice cream in a bag'. The mixture undergoes a change by shaking the sealed bag of heavy cream, milk, sugar, and vanilla extract, as the shaking causes the ingredients to mix. The ice and rock salt in the outer bag cause the mixture's temperature to drop, with the mixture undergoing a chemical change as it turns into ice cream.

Materials:

- 1 large ziplock bag

- 1 medium ziplock bag
- 1 cup heavy cream
- 1/2 cup sugar
- 1 teaspoon vanilla extract
- 1 cup ice
- 1/2 cup rock salt
- a small cup of crushed fruit or chocolate chips (optional)

Instructions:

1. Combine the heavy cream, sugar, and vanilla extract in the medium ziplock bag. Seal the bag tightly, making sure there are no leaks.

2. Fill the large ziplock bag with ice and add the rock salt. Seal the bag tightly.

3. Place the medium lock bag containing the cream mixture inside the large ziplock bag with the ice and salt.

4. Shake the bags vigorously for about 10-15 minutes until the mixture thickens and becomes the consistency of soft-serve ice cream.

5. If desired, add a small cup of crushed fruit or chocolate chips to the mixture.

6. Serve the ice cream in small cups or cones, and enjoy!

Note: Make sure to supervise students while shaking the bags, as the salt can cause the bags to become very cold and potentially painful to touch. Wearing gloves or using a towel to hold the bags is recommended.

Discussion Questions:

- Can you explain how the liquid cream mixture is transformed into a solid ice cream through the process of freezing?

- Can you explain how adding different ingredients such as fruit, chocolate chips, or extracts can alter the texture and flavour of the ice cream and how this relates to the chemistry of the ingredients?

- Can you explain how the salt lowers the freezing point of the ice, allowing the cream mixture to freeze at a lower temperature?

Candy Crystals

Context: This is a sweet activity that students can do when exploring the concepts of solutions and mixtures. In this experiment, students will be able to visually see how a solution changes as it moves from being dilute to being saturated while also including the concept of states of matter by being able to see the saturated solution evaporate to reveal the yummy candy crystals.

Materials:

- Plastic cups or glass jars
- Water
- Sugar

- String
- Paddle Pop stick or pencil
- Kettle or hot plate
- Clean bead or paper clip
- Coffee filter or paper towel

Instructions:

1. Students tie one end of the string to the middle of the Paddle Pop stick or pencil and tie the other end around a bead or paper clip.

2. Lay the stick or pencil across the top of a jar so that the string hangs down the middle. Ensure that the string does not touch the bottom of the jar, but it should be close. Also, ensure that it is not touching the sides either.

3. Boil 1 cup of water in the kettle or over a hot plate.

4. Add 1/4 cup of sugar to the boiling water and stir until it dissolves.

5. Repeat this step until all of the sugar has dissolved. This will take time and patience, and it will take longer for the sugar to dissolve each time.

6. Carefully pour the hot sugar solution into the jar and fill it almost to the top. Place the Paddle Pop stick or pencil back on top and lower the bead or paperclip back inside the jar. Place a coffee filter or paper towel over the jar so nothing falls in.

7. Allow the jar to cool and put it someplace where it will not be disturbed. Check on it every day to see the crystals start to grow, but be very careful not to disturb it. The longer you wait, the bigger the crystals will get.

Discussion Questions:

- How does the process of crystallisation work in this experiment, and how is it related to the properties of saturated sugar solutions?

- How does the speed of the crystallisation process change the size of the crystals that form? Why is this the case?

- What is the purpose of the bead or paperclip that is attached to the string in the jar? Would you still have crystals form if this was missing?

Elements 'Wanted' Poster

Context: In this activity, students make a 'wanted' poster highlighting the properties of their chosen element. Students will be required to research their element to find information regarding a range of chemical and physical properties and display this in an interesting and engaging way. Depending on your class size, this activity could be used to create a collaborative class periodic table. QR codes could also be incorporated into the posters to provide an interactive element to the activity.

Materials:

- Poster-making materials
- Internet

Instructions:

1. Students are provided with an element to research.

2. Students use resources to find information regarding the element's physical and chemical properties.

3. Students create wanted posters and display them in the classroom alongside their classmates' posters to create a collaborative periodic table.

Elephant's Toothpaste

Context: Elephant's toothpaste will always get a reaction... literally! It is a great way to demonstrate several different scientific concepts, including chemical changes, rates of reaction, endothermic vs ectothermic reactions, and even the use of catalysts in chemistry. This experiment can be done as a demonstration or as a student activity. The great thing about doing this as a demonstration is that you can really make it an experience that your students won't forget... trust me!

Materials:

- 500 mL measuring cylinder
- 100 mL 30% hydrogen peroxide solution
- Potassium iodine powder

- Tablespoon
- Dishwashing liquid
- Food colouring
- Splint
- Matches
- Disposable gloves
- Safety goggles

Instructions:

1. Set up the equipment outside so that when it makes a mess, it won't ruin your classroom.

2. Pour 50 - 100 mL of 30% hydrogen peroxide solution into the measuring cylinder.

3. Squirt in dishwashing liquid and swirl it around.

4. Place 5-10 drops of food colouring along the wall of the cylinder to make the foam resemble striped toothpaste.

5. Add a large tablespoon of potassium iodine powder. Do not lean over the cylinder when you do this, as the reaction is very vigorous, and you may get splashed or burned by steam.

6. Once the reaction has slowed down, touch a glowing splint to the foam to relight it. This indicates the presence of oxygen in the foam. You may also allow students to come past and 'feel' the warmth of the foam without touching it directly.

7. Clean up the mess!

Discussion Questions:

- What chemical reaction produces the foam in the elephant's toothpaste experiment?

- How would changing the concentration or temperature of the hydrogen peroxide affect the size and duration of the foam produced in the experiment? Can you explain why these changes occur?

- Can you explain the role of potassium iodide as a catalyst in the reaction and how this relates to biological processes in the body?

Firewriting

Context: Firewriting is a fun and easy chemistry demonstration involving the combustion of metal salts. It is a popular demonstration used in chemistry classes to engage students and illustrate combustion, energy, and spectroscopy principles. By dissolving the metal salts in water, secret messages can be 'painted' onto absorbent paper and then lit alight to reveal the message to students.

Materials:

- Sodium nitrate saturated solution
- Absorbent paper, such as sugar paper
- Small paintbrush
- Wooden splint
- Matches

Instructions:

1. Using the sodium nitrate solution, use the small paintbrush to 'paint' a secret message onto the sugar paper. Ensure that a thick layer of solution is used.

2. Allow the paper to dry completely. This can be done naturally, or you can speed up the process with a hair dryer.

3. Light a wooden splint, then gently blow it out so the end just glows.

4. Touch the glowing end to the start of the secret message until the treated paper starts to glow and char. The paper may need to be tilted so that the charring continues along the painted line.

Safety considerations:

Care must be taken so the paper doesn't burn too vigorously – it can burst into flames unexpectedly.

As this reaction produces a lot of smelly smoke, it is important to take care when disposing of burnt and unburnt papers and to ensure good ventilation in the lab.

Discussion Questions:

- How does the process of combustion work, and how is it related to the fire writing activity?

- Can you explain the chemical reactions that occur when the solution is ignited and how they produce light and heat?

- Can you explain the importance of proper ventilation, handling of potentially hazardous chemicals, and use of personal protective equipment such as gloves and safety glasses?

Periodic Table Battleship

Context: This is a fun activity that can be scaled up or down in difficulty based on the ages of the students playing. The goal of playing this game is to have students develop a deeper understanding of the location of elements within the periodic table and become comfortable with naming and locating them easily.

Materials:

- 4 copies of the periodic table
- Laminator
- 2 manila folders
- Paper clips
- 2 whiteboard markers

Instructions:

1. Laminate four copies of the periodic table and stick them inside the two manila folders.

2. In pairs, students will use the paper clips to hold the two manila folders upright to look like a battleship game.

3. Using the whiteboard markers, the students will mark out a number of 'ships' on their periodic table.

THE ENGAGING SCIENCE CLASSROOM

Depending on the age of the students, this can be made easier or more difficult. For example, for more senior Chemistry students, setting requirements around the location of their ships will ensure that they are using their understanding of the periodic table along with the game's fun. For example, one ship needs to be located within the transition metals and one in the alkali metals.

4. Students are to then call out coordinates (rows and groups), element names, atomic numbers, or mass numbers to 'sink' their opponent's ship. Students use their second periodic table to keep track of the places they have 'dropped' to keep track of their hits and misses.

5. The student who sinks all of their opponent's ships is the winner.

Modelling Ionic Structures

Context: In this activity, students will use lollies and toothpicks to model different ionic structures. This activity could also be carried out using Molymod kits or plasticine; however, adding lollies to the mix will help students be more engaged.

Materials:

- A variety of lollies to represent atoms
- Toothpicks to represent bonds

Instructions:

1. Divide the students into small groups and provide each group with a set of lollies and toothpicks. Explain to the students that the lollies represent atoms and that the toothpicks represent the bonds that can form between them.

2. Have the students use the lollies and toothpicks to model the structure of different ionic compounds. For example, they might use red lollies to represent atoms with a positive charge and blue lollies to represent atoms with a negative charge.

3. As the students build their ionic structures, have them consider the following:

   a. The number of ions present in the compound

   b. The charge of each ion

   c. The arrangement of the ions in the compound

4. At the end of the activity, debrief with the students about what they observed during the activity. Ask them to consider the characteristics of ionic bonds and how they contribute to the structure and properties of ionic compounds.

Ionic Formula Cubes

Context: In this activity, students will gain confidence in understanding how to create chemical formulas, including knowledge of ionic bonding. This activity can be scaled depending on the students in the group and used for simple chemistry concepts or more complex concepts such as polyatomic ions.

Materials:

- A set of 'cation' cardboard cubes printed in one colour
- A set of 'anion' cardboard cubes printed in another colour
- Sticky tape

Instructions:

1. Create the cation and anion cubes by writing the symbol and valency of 6 different cations on one cardboard cube and six different anions on a second cube. Placing tape on all edges will ensure that the cubes last longer. Students could also make these cubes to identify ions and their valencies.

2. Provide each group with a set of 3 cation and 3 anion cubes.

3. Students are to 'toss' one of each cube and make note of the name and symbol of each ion, then determine the chemical formula of the resulting ionic compound.

4. Students repeat this a number of times until they have mastered the concepts.

Half-Lives with Skittles/M&Ms

Context: In this experiment, students will use Skittles or M&Ms to represent radioactive atoms to better understand the concept of half-life. Students will shake the Skittles and pour them out onto a paper plate. Those that land with the blank side up are not radioactive and will therefore be known as the element Blankium (Bl). Skittles that land with their 'S' side up will be considered radioactive, known as the element Skittilium (Sk).

Materials:

- Paper cup with Skittles or M&Ms
- Paper plate
- A3 paper to cover the desk
- Graph paper

Instructions:

1. Pour the Skittles onto the plate so that none of them touch the desk. Count out the Skittles and place them back in the cup. Record the exact number of Skittles you are starting with.

2. Place a hand over the cup, shake for 10 seconds (the half-life for Skittles), and then pour the Skittles onto the plate. Do not turn any over! This represents one half-life or decay period. Count all of the non-radioactive Blankium atoms and record the number.

3. Count the number of radioactive Skittilium atoms remaining and record the number in a table. Put them back in the cup. Shake them for 10 seconds and pour them back onto the plate. This represents a second half-life.

4. Again, remove the Blankium atoms that have decayed, record how many are left, and then eat them. Count the radioactive Skittilium atoms that remain and record this number.

5. Repeat these steps until all of your Skittles have decayed or your radioactive Skittles have gone through 8 half-lives.

Discussion Questions:

- How does the process of radioactive decay work, and how does it relate to the activity using Skittles?

- Can you explain how the half-life of a radioactive isotope is used to determine the safety and effectiveness of medical treatments or the stability of nuclear power plants?

- What factors can affect the decay rate of radioactive isotopes, and how can these factors be manipulated in the Skittles activity?

Balloon Combustion

Context: This is only to be done as a demonstration due to the dangerous nature of the activity. This demonstration is a great way to introduce the concept of combustion to students, as it is loud and bright! Have the balloons prepared before the class so they are ready to go. It is best to have a spare of each. The balloon with helium will simply pop, while the balloon with hydrogen will ignite with a loud bang and a bright flash.

Materials:

- Balloon filled with helium
- Balloon filled with hydrogen
- Candle attached to the end of a metre ruler
- Matches
- Safety goggles

Instructions:

1. Ask the students to make their observations of the two balloons and to write down what they think is in the balloons - they should be fairly similar as they will both be floating.

2. Light the candle using a match and hold it next to the first (helium-filled) balloon until it pops. Have the students write down their observations of what took place, including what happened to the candle when the balloon popped.

3. Light the candle again and hold it near the second balloon. This time, it will pop with a much louder sound and a bright flash. Have the students repeat their observations and compare the differences between the two balloons.

4. Engage in a class discussion about what the students thought was in the balloons and why they experienced what they saw and heard.

Discussion Questions:

- Can you explain why the balloon filled with hydrogen gas is more combustible than the one filled with helium gas?

- What safety precautions should be taken when working with combustible gases like hydrogen?

- How can the concept of combustion be applied to real-world scenarios, such as in the field of rocket science or energy production?

BIOLOGY ACTIVITIES

Colourful Flowers

Context: In this activity, students will use flowers to observe the movement of water through a plant. White roses and carnations are the best flowers for this experiment, giving the best results and the brightest colour changes. The use of concentrated food dye is recommended to get the greatest colour change. Students can also split the flower's stem and place each piece of the stem into different coloured food dye to create a rainbow effect. This method requires using a scalpel or sharp knife to split the stem, so it is recommended for older students. It is also important that the stem is not broken when it is split, or the water cannot travel through to the petals.

Materials:

- A white rose or carnation

- Small jar
- Food colouring
- Water
- Scalpel or knife

Instructions:

1. Place food colouring and water into a small jar.
2. Cut the bottom of the stem approximately 5 cm from the end.
3. Place the flower into the jar and leave it on a windowsill.
4. Observe the changes to the flower petals over time and note the colour changes.

Discussion Questions:

- How does the process of water transportation work in plants, and how is it related to this experiment using flowers and coloured water?
- Can you explain how water's cohesive and adhesive properties allow it to form a continuous column within the xylem vessels and how this allows it to move from the roots to the leaves?
- Can you explain how changing the temperature or humidity of the environment the flower is kept in can affect the rate and efficiency of water transport? How does this relate to the adaptations of different plant species to different environments?

Ecosystems String Activity

Context: This activity is a great way for students to see how everything in an ecosystem interacts and how various external factors can impact an entire ecosystem. It requires a large area where the students can spread out and a ball of string that can be unravelled as the ecosystem connections occur. Before this activity, students must know about the various elements that make up an ecosystem. They may also need to research different ecosystems to be able to create their name cards before they complete the string activity.

Materials:

- Half a sheet of A4 paper per student
- Pen
- Ball of string

Instructions:

1. As a class, decide on an ecosystem to model, such as the Great Barrier Reef, Simpson Desert, or Daintree Rainforest.

2. Each student is to select something to be from the ecosystem, e.g. sun, grass, fish, cat, soil. Each person must be different and create a label on a blank paper sheet. If needed, the teacher can have these sheets pre-prepared to ensure enough variety and potentially save time.

3. Have all students stand in a circle, holding their sign up so everyone can see.

4. Choose one person to hold the end of a ball of string. That person then needs to throw the ball (carefully!) to a thing in the environment that they interact with (to another student), e.g., a fish can throw the string to the water, and water can throw the string to the tree.

5. When throwing the ball of string, students should state what the relationship or interaction is, e.g. when the fish throws the string to the water, the fish states, 'I need water to live in', and the water then states, 'I give water to trees' and throws the string to the trees and so on.

6. Once the ball has been passed to all students, ask the class to discuss the 'web' they have created and how all things within an ecosystem are related.

7. Announce to the class that one of the organisms within the ecosystem has been impacted by human activity (such as pesticides or deforestation) and that the student with this organism needs to drop their string and move out of the circle. Ask the student what happens to the organisms immediately on either side of this organism. For example, what would happen to the number of producers in the ecosystem if this was a first-order consumer? Or what would happen to the number of animals that directly eat that animal or those further up the food chain?

8. If time permits, have the original student stand up and rejoin the web before repeating the activity with another 'organism'.

Discussion Questions:

- How many different webs could be created with this activity?

- Think about how different human activities can cause different impacts on an ecosystem depending on the organisms directly affected. How would a terrestrial impact differ from an aquatic one?

Modelling Bioaccumulation

Context: In this activity, students will model how bioaccumulation can impact an ecosystem. Lollies will represent an algal bloom that has impacted a food source in a waterway, and students will represent the different animals within the ecosystem. Before beginning, divide the class up into groups - approximately 60% of the students will be the krill, 25% fish, 14% seals, and 1% of the students will represent the killer whales. Students will be given a lunch bag to collect their 'food' by following the instructions below. You should also determine which lollies will be 'toxic' at the start of the activity.

Materials:

- A bag of mixed lollies (may be wrapped so that they can be eaten or placed on a tarp to avoid contamination)

- A lunch bag for each student

Instructions:

- Spread the lollies out in an open area and ask only the 'krill' to begin collecting them. Tell the students who are the krill that they will have 30-60 seconds (depending on the number of students involved and the size of the area) to graze. They should move around the area feeding on the algae (collecting lollies and placing them in their food bags). Remind the students to refrain from actually eating their food since they will need to assess their feeding success at the end of the game. At the end of the timed period, the krill should remain where they are in the area but stop collecting lollies.

- Tell the students who are the fish to enter the feeding area. Tell them, 'You have come across a swarm of krill feeding on an algal bloom, and you have 20 seconds to feast on the krill', and 'Krill, you may now continue feeding'. In order to simulate eating the krill, the students who are the fish take the food bags from the students who are the krill and empty their lollies into their bags. The fish cannot 'eat' another krill until all the lollies are placed into their bag. Once the krill has been 'eaten', they must sit out.

- All the remaining krill must now sit out. The students who represent the seals have 15 seconds to feed on the fish following the same instructions as when the fish ate the krill.

- All the remaining fish must now sit out. The students who represent the whales have 5 seconds to feed on the seals, once again following the same instructions.

- Once finished, the students count the number of lollies that they have left in their bags. Let the students know which of the lollies were 'toxic' and determine which animal has the most of these in their food bag.

Discussion Questions:

- How does this activity model the process of bioaccumulation in living organisms, and what are the key factors involved?

- Can you explain how chemicals and pollutants enter the food chain and accumulate in the tissues of organisms and how this can negatively impact the health and survival of those organisms?

- Draw a diagram of the ecosystem's food chain and show how the toxic algae was passed from one trophic level to the next.

Classification Activity

Context: In this activity, students will understand how using classification techniques helps identify everyday objects more easily. They will then be able to transfer this knowledge to the classification of plants and animals as they continue to work through a unit on how and why we classify in science.

Materials:

- A tray or box
- A tea towel or cloth
- A stopwatch
- A range of random household items that can be grouped based on characteristics

Instructions:

1. Have the tray of items set up in the room with the cloth covering it before the students walk into the room to create some interest in the activity.

2. Ask the students to stand around the tray and be prepared to remember as many items as they can in 30 seconds. Students cannot write anything down or talk with their peers.

3. Lift the cloth and start the timer. After 30 seconds, return the cloth to the tray and ask the students to write down as many items as possible on scrap paper. Collect the papers so students cannot refer to them again.

4. Discuss classification and using characteristics to group items together with the students.

5. Remove the cloth a second time, and start the timer. This time, the students are encouraged to group things together to help them remember the items.

6. After 30 seconds, return the cloth and have the students write down as many items as possible.

7. Ask the students whether they were able to remember more items by using the methods of grouping items together. Discuss with the class the ways students grouped items together.

8. For the last time, remove the cloth, have students group all items into smaller groups, and begin discussions around creating keys for classification.

9. Have the students create a branched dichotomous key for the items that were on their tray.

Pokemon Go Classification

Context: In this activity, students will use Pokemon Go characters to create dichotomous keys to classify the different characters in the game. This activity can be done with any character from a TV show or movie or any other group of characters that students are interested in. Students will develop the skills to build branching and tabular dichotomous keys.

Materials:

- Printouts of Pokemon Go character cards
- A3 paper
- Glue
- Ruler
- Markers

Instructions:

1. Students are given a set of cards and must create a branching key to classify each Pokemon character.

2. Students paste the cards at the end of each branch to finalise their branching key.

3. Students use their branching key to create a tabular key to classify each character.

4. Students compare dichotomous keys with other groups to see what characteristics were used to classify the Pokemon characters.

Ziplock Digestion

Context: In this activity, students will be able to observe the processes of mechanical and chemical digestion of food. By placing the items into a ziplock bag, students can easily see what happens at each stage and compare these to what occurs inside the body when we eat and digest our food. This also gives students an opportunity to discuss the benefits and limitations of models such as this.

Materials:

- Ziplock bag
- Milk Arrowroot biscuit (or similar)
- Specimen jar with 20-30 mL of lemonade
- Paper plate

Instructions:

1. Take a photo of the biscuit and note its size, shape, and consistency.

2. Place the biscuit into the ziplock bag and seal it, ensuring no air is inside the bag.

3. Carefully mash the bag, noting what happens to the biscuit. The biscuit should turn into smaller pieces.

4. Carefully open the bag, add the lemonade, and seal it again, ensuring there is no air in it.

5. Roll the bag around carefully while observing any further changes to the biscuit.

6. Open the bag and slowly pour the mixture onto the paper plate. Take a photo of the final result and note the difference between the biscuit at the beginning and now.

Discussion Questions:

- Identify the structures involved in digestion that are modelled in this activity.

- Compare the consistency of the items placed in the bag at the beginning of the activity with the consistency at the end of the activity.

- How does this activity model the process of food digestion in the human body, and what are the key steps involved? What limitations of this model exist in demonstrating the processes that occur during digestion?

Pathways Around the Body

Context: This activity can be used to demonstrate the pathway of substances around a plant or animal. For example, students could follow the pathway of blood around the body, including through the lungs, stomach and digestive system, demonstrating what items are delivered to and collected from different locations on their journey. Another example could be the pathway of substances through the kidney's nephron, as this is difficult for students to understand.

Materials:

- A large space where the pathway can be mapped out on the floor
- Cards with substances printed on them - e.g. carbon dioxide, oxygen, glucose, etc
- Cards with organs printed on them - e.g. lungs, stomach, large intestines
- A video to show the pathway for students to visualise the process before beginning

Instructions:

1. Draw out the structures that make up the pathway on a large space that will allow the students to move around.

2. Hand out cards to different students to have them act as different substances or organs in the body. Have

students move to their location on the drawn-out map of the body.

3. Start the process by choosing where the substances will begin. Students who are the relevant substances are to move around the pathway and interact with their appropriate organs.

4. Complete the activity a number of times with students taking on different roles so that all students have a turn at being involved in the activity.

Osmosis with Gummy Bears

Context: In this experiment, students will use gummy bears to model osmosis. By understanding osmosis, we can learn about the movement of water and how it affects cells, and the gummy bears provide a great visual representation of this action. Students will observe that the gummy bear in the plain water remained the same size, while the gummy bear in the salt solution shrank and the gummy bear in the sugar solution grew.

Materials:

- Gummy bears
- Distilled water
- Concentrated salt solution
- Concentrated sugar solution
- 3 small cups or containers

Instructions:

1. Place a gummy bear in each cup.

2. In the first cup, add plain water to cover the gummy bear. Add a salt solution to cover the gummy bear in the second cup. Add a sugar solution to cover the gummy bear in the third cup.

3. Observe the gummy bears over the course of an hour, noting any changes in size or shape. If no significant change exists, leave the gummy bears for longer.

4. After an hour, remove the gummy bears from the cups or containers and compare their sizes.

Discussion Questions:

- Identify the cellular structures that are represented by the gummy bear and the different solutions in this experiment.

- Explain how water molecules move from an area of high concentration to an area of low concentration through a semipermeable membrane and how this affects the size and shape of the gummy bears.

- Explain how solutions with different concentrations of solutes can affect the movement of water across cell membranes and how this can affect the function of cells and tissues.

Neuron Cookies

Context: In this activity, students will use a biscuit and lollies to model the structure of a neuron. Before building their model, students must understand the parts of the neuron and how each of the various items in their ziplock bag may represent these structures. Students will then use their understanding of the structure to recreate the neuron using these items before labelling each part and taking a photo of it (before eating it, of course!).

Materials:

- Ziplock bag containing the following: a round or oval-shaped biscuit, mini marshmallows, confectionary snake, Strawberry & Cream (or other oval-shaped lolly), raspberry licorice straps (or similar)
- Paper plate
- Pre-made icing
- Knife
- Textas or coloured pencils

Method:

1. Research the structure of a neuron and assign each piece in the ziplock bag a part of the model that it will represent.
2. Create a model of the neuron on the paper plate.
3. Label all parts appropriately and note their function.
4. Take a photo of the labelled model.

Pineapples & Jelly Enzymes

Context: In this experiment, students will investigate how the process of heating pineapple during the tinning process denatures the enzymes. Students will compare how jelly sets when both fresh and tinned pineapple are placed in the mixture before being placed into the fridge to set. The fresh pineapple will not set properly as the enzymes are still active and prevent the gelatin from forming, while the tinned pineapple will set as these enzymes are no longer functioning.

Materials:

- Tinned pineapple cut into 1cm cubes
- Fresh pineapple cut into 1cm cubes
- Packet jelly
- Boiling water
- Plastic bowl
- Spoon
- Clear jars to set the jelly and pineapple mix

Instructions:

1. Prepare the jelly as per the packet instructions.
2. Label one plastic cup 'Fresh', one 'Tinned', and one 'Control.
3. Add fresh-cut pineapple and 200 mL of jelly to the' Fresh' cup.

4. Add the same amount of tinned pineapple and 200 mL of jelly to the' Tinned' cup.

5. To the 'Control' cup, add 200 mL of jelly.

6. Place the three cups into a fridge to allow the jelly to set.

7. Remove from the fridge and observe the jelly in the three cups.

Discussion Questions:

- Which of the three jellies did not set? Which pineapple was placed into this jelly?

- Bromelain is an enzyme in pineapples that causes gelatin to denature. Use this information and your understanding of proteins to explain why fresh pineapples cause jelly not to set.

- Explain the process of canning and how this impacts the outcome of this experiment.

Modelling Disease Transmission

Context: There has been a lot of news about infectious diseases throughout recent years. In this class activity, the students will model how an infectious disease is transmitted through a population. A 'basic' solution is placed in one test tube while the remainder of the test tubes contain deionised or distilled water. Students are asked to exchange their solutions with others; this is designed to mimic disease transmission. Once students have exchanged solutions, a Phenolphthalein indicator solution is used to reveal which students have become 'infected' with the simulated virus. Being able to identify the original carrier of a disease is a critical skill for epidemiologists. An epidemiologist will work backward to identify the viral source when faced with epidemic viral infections. Ask students to consider how finding the 'original carrier' in your class population compares to real cases of global epidemics.

Materials:

- 10 numbered test tubes - 9 containing 5 mL of distilled water, 1 containing 5 mL of 0.2 M sodium hydroxide (ensure you remember which test tube is the 'carrier' with the sodium hydroxide)
- 10 numbered pipettes
- Phenolphthalein solution

Instructions:

Spreading the infection

1. Each student collects a numbered test tube and a pipette (of the same number) and notes the number in their records.

2. Students move around the room and find a partner. Each student is to record the other's name and number in their records.

3. One student will use the pipette to draw up roughly half of the solution in their test tube and transfer it to their partner's test tube. Their partner will now transfer half of their solution to the original student's

test tube. Students should mix the contents of the test tube by carefully stirring with their pipettes.

4. Students will repeat these steps twice more while moving throughout the classroom. Ensure the students do not exchange fluids with the same person twice.

5. After three transfers, add 2 drops of Phenolphthalein solution to each test tube.

6. Identify which students' solutions carry the virus. If the solution remains clear or turns yellow, the solution is negative for the 'virus'. If the solution turns red or pink, the solution is positive for the 'virus'.

Finding the carrier

- Have the students who tested negative write the number of their test tubes on the board. Students should cross out the 'negative' numbers and do not need to include their number, as they cannot possibly be the original 'carrier'.

- Ask the 'positive' students who exchanged solutions with a 'negative' student to identify themselves. Students should cross out the numbers of these 'positive' students from the possible 'carrier' list and add their numbers to the 'negative' list.

- After this, the number of 1 or 2 students should remain. These are the final possible original carriers. When two remain, one is the original carrier, and the other is the first infected student.

Discussion Questions:

- How does this experiment represent the transmission of disease in a real-life context?

- What challenges did you face in identifying the original carrier, and how might these be reflected in real life?

- What are some ways that we can minimise the spread of infection?

- Thinking about the recent COVID-19 pandemic, what methods of minimising disease transmission could have been used in this activity? What methods would not work in this case?

- Why must epidemiologists work as quickly as possible to determine the origin of an infectious disease?

Extracting DNA From Strawberries

Context: This activity allows students to extract DNA from strawberries using basic items found around the house. Although not a difficult activity to carry out, this activity brings out a sense of wonder and awe in your students when they see the 'snotty' DNA that they can extract. Each step plays an important part in the process of extracting the DNA. Crushing the strawberries in the zip lock bag helps to break open many of the strawberry cells, where the DNA is. The dishwashing detergent breaks down the membranes of the cells, releasing the DNA. The salt makes the DNA molecules stick together and separate from the proteins that are also released from the cells. The coffee filter will retain cell debris and unmashed pieces of fruit. The DNA will pass through the filter into the glass. DNA is not soluble in alcohol, so it precipitates. What you see are long, ropelike DNA molecules in the alcohol.

Materials:

- 2-5 strawberries (fresh or frozen)
- Ziplock bag
- 2 x 250 mL beakers
- 10 mL measuring cylinder
- 5 mL dishwashing detergent
- Teaspoon
- Teaspoon of table salt
- 250 mL measuring cylinder
- 125 mL water
- Spatula
- Coffee filter
- Rubbing alcohol
- Metal paperclip

Instructions:

1. Discard any stalks that remain on the strawberries and place the strawberries into the ziplock bag. Close the bag, ensuring that there is no air trapped inside.

2. Crush the strawberries by gently 'smashing' them in the bag until they are completely crushed. This starts to open the cells to release the DNA.

3. Mix 5 mL of detergent, one teaspoon of table salt, and 125 mL of water in one of the beakers to make your DNA extraction liquid.

4. Add approximately 10 mL of the DNA extraction liquid to the bag of strawberries. This will further break open the cells. Reseal the bag and gently smash it for another minute. Avoid making too many soap bubbles.

5. Place a coffee filter inside the opening of the other beaker. Open the ziplock bag and pour the strawberry liquid into the filter. You can carefully twist the coffee filter just above the liquid and gently squeeze the remaining liquid into the beaker.

6. Remove the coffee filter from the beaker and set it aside.

7. Pour down the side of the beaker the same amount of cold rubbing alcohol as there is strawberry liquid. The alcohol should float on top of the strawberry liquid. DO NOT mix or stir.

8. Watch the beaker for a few minutes, looking for a white, threadlike substance to rise out of the strawberry liquid and into the alcohol layer. This substance is DNA.

9. Unfold a paperclip and bend the end to give it a hook. Insert the paperclip into the DNA and twist

it to collect it. Slowly lift the collected DNA out of the solution.

10. Handle the DNA to see what it feels like. Record any observations.

Discussion Questions:

- Can you explain how the fruit cells are broken down to release the DNA and how the DNA is separated from other cellular components using household items such as dish soap and alcohol?

- Why do different fruits or vegetables contain different amounts and types of DNA? Could this experiment be used to study genetic diversity in plants?

- How is DNA extraction used in fields such as forensic science and biotechnology?

Genetic Predictor of Doom

Context: This activity is a great way to demonstrate how genetic characteristics are passed from generation to generation. Students often struggle with the fact that with recessive genetic traits, there is a 25% chance of the trait being passed on to each child of two healthy parents rather than just a 25% chance of the trait being passed on to all children a couple has. This lesson can be delivered using any genetic trait that is passed on with a 25% chance, so if you think that the distressing stories of children passing away from Tay-Sachs may be triggering for your students, choose another genetic trait, such as albinism. A huge thank you to Chris Young (@chtyoung on Twitter) for sharing this lesson on Twitter so many years ago!

Materials:

- A large box with a removable lid - an empty A4 paper box works perfectly - with four holes cut into the side of the box along the bottom - like four 'mouse doors'

- Four spring-loaded mouse traps taped inside the box, lined up with the holes

- Ruler or stick

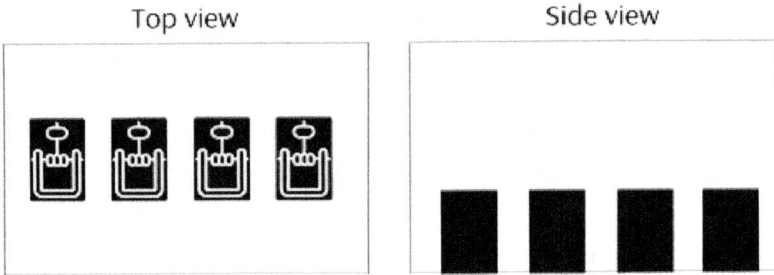

Top view   Side view

Instructions:

- Begin the lesson with a question for your students - 'If two healthy parents have a child with Tay-Sachs disease, what are the odds their next child will have the same disease?' Provide them with multiple choices - 0%, 25%, etc.

- After the students have thought about their answer, begin a discussion about Tay-Sachs disease, including its impact on the body, how it is diagnosed, and the usual outcome for those diagnosed. You may even find a story about a child who was diagnosed with Tay-Sachs at a young age that explains the impacts that this disease had on the child and their family. This will help to get your students invested in the lesson.

- Pose another question to your students - 'If your first child was diagnosed with Tay-Sachs disease, would you have another child?' Ask them to think about the risks of having another child when they are unsure of the actual genetic probability of passing on the disease. Explain to the students that they can never truly feel the pain of the parents of a child with Tay-Sachs disease but that you'd like to help them.

- Bring out the 'Genetic Predictor of Doom' and place it on the desk without opening it or revealing what is inside.

- Ask for a volunteer to be the 'parent' and send them out of the classroom. Open the box and reveal to the class what is in it. Set one mousetrap, close the box carefully, and bring the volunteer back into the classroom.

- Without telling the volunteer what is in the box, allow them the chance to declare where they are putting their finger - this represents their child's genetic outcome. Do not let them put their actual finger in the trigger hole!!

- Give the volunteer a ruler or stick to place directly into the hole. If they set the mouse trap off, then they had a child with Tay-Sachs. If they did not, then they had a healthy baby.

- Reset the mouse trap (setting a different one each time) to let each class member have a go. Record your results; if the probability is correct, it should be about 1 in 4 or 25%.

Discussion Questions

- What is the difference between the inheritance of dominant and recessive traits?

- How could the original parents be healthy despite having a child with Tay-Sachs?

- Draw a Punnett square to check that the probability that was calculated by the class was correct.

Gummy Worm Karyotypes

Context: In this experiment, students will use gummy worms to model chromosomes to understand how karyotypes are used to identify genetic disorders. By creating a karyotype with gummy worms, students can learn about the structure and function of chromosomes and how they can be used to diagnose genetic disorders.

Materials:

- Gummy worms
- Scissors
- Coloured pencils or markers
- A sheet of paper

Instructions:

1. Cut the gummy worms into small pieces that represent chromosomes. Each gummy worm should represent one chromosome pair, so you should have a total of 46 pieces.

2. Using coloured pencils or markers, label each piece of gummy worm with the corresponding chromosome number. For example, you should label one piece '1' and another piece '2' to represent the first chromosome pair.

3. Arrange the gummy worm pieces on the sheet of paper to create a karyotype. Make sure to place the chromosomes in order based on size and shape.

Discussion Questions:

- Using what you know about genetics and karyotypes, identify any genetic disorders that may be present based on the arrangement of the gummy worms. Use the internet to research the disorder and identify the impact this disorder has on an individual.

- Explain how karyotypes can be used in real-life contexts.

Jellybean Natural Selection

Context: In this activity, students will use jellybeans of different flavours to model natural selection, where the preferred flavour becomes extinct quickly. Students will eat one of each jellybean of their preferred flavour until only six remain in their pair. Students will combine their results to determine which flavour was selected (the one with the most remaining) and discuss why this may be the case.

Materials:

- 20 jelly beans per pair of students
- Stopwatch

Instructions:

1. In pairs, students lay down a sheet of paper and place the jellybeans on top of the paper.

2. Every 30 seconds, students are to eat a jellybean until six jellybeans are left. Students are to note the colours of the jellybeans that remain.

3. Students pool their results on the board for the different colours of jellybeans remaining. There should be a dominant colour of jellybeans remaining, e.g. black. In nature, this could be represented by a poisonous species that predators do not attack.

Discussion Questions:

- How do variations in traits, in this case, the flavour of jellybeans, lead to changes in the number of organisms with different traits within a population over time?

- In this population of jellybeans, what will happen to the population of each jellybean colour with each generation?

- Explain how changes in the environment, such as the introduction of a new predator or the availability of different food sources, can influence the survival and reproduction of individuals with different traits and how this can lead to the evolution of new traits in a population.

Colour and Survival

Context: This activity has students model the benefits of physical adaptations when hiding from predators. This is a simple activity to set up and carry out with any number of students in any classroom. This activity can be done outside with green toothpicks on the grass, or an artificial environment can be created inside the classroom using a range of different materials. Students will take turns being the predator, searching for food!

Materials:

- Rope
- Coloured toothpicks (ensure a percentage are the same colour as the chosen 'environment/s')
- Stopwatch
- Different coloured environments - this can be grass or shredded paper, for example

Instructions:

1. Rope off 1 square metre of the chosen 'environment'.

2. Have one student turn their back while others scatter various-coloured toothpicks.

3. The single student turns around and is given 30 seconds to pick up as many toothpicks as possible (one toothpick at a time – using only two fingers – imitating a bird picking up an insect)

4. At the end of 30 seconds, the other students count the numbers of each colour that has been gathered.

5. Record the number of toothpicks of each colour.

6. Students repeat this for all the group members and for all 'environments'.

7. Find the average number of toothpicks collected of each colour in each environment.

Bubble Membranes

Context: It can be difficult for students to visualise the processes that occur in the cell membrane. In this activity, students will use bubbles to explore some of the more complex concepts about how the cell membrane works. Bubbles make a great stand-in for cell membranes as they're fluid, flexible, and can self-repair. Bubbles and cell membranes are alike because their parts are so similar.

Materials:

- Shallow dish
- Bubble mixture - dish soap with a small amount of water

- 'Bubble frame' made of 4 bendable straws or wire coat hanger
- 1 clean straw
- Small string loop
- Large string loop

Instructions:

a. Exploring the fluidity of the cell membrane

1. Place the bubble frame in the shallow dish and add bubble mixture to slightly cover it.
2. Lift the bubble frame out of the solution so that a thin film spaces across the frame.
3. Tilt the frame back and forth and observe the surface of the film.
4. Notice the swirl of colour as the light reflects off the film. Molecules in the cell membrane move about in a similar fashion.
5. Hold the frame by the edges and rotate the sides in opposite directions. Notice the elasticity of the film.
6. Hold the bubble frame parallel to the floor and gently move the frame up and down until the surface bounces up and down.

b. Modelling the way membranes can self-repair

1. Lift the bubble frame out of the solution so a thin frame spaces across the frame.

2. Cover the surface of your finger or an extra straw in the bubble solution.

3. Slowly push your finger or the straw through the film. The film should allow it to pass through without breaking.

4. Remove your finger or the straw; the film should repair itself.

5. Try the same procedure with your entire hand.

c. Modelling membrane-bound organelles

1. Place the tip of a clean straw into the bubble solution in the tray. Gently blow on the other end of the straw to create a bubble.

2. Slowly lift the tip of the straw out of the liquid while continuing to fill the bubble with air. Allow the bubble to grow to a size of about 10 cm wide.

3. Return the tip of the straw to the bubble solution and try to create a smaller bubble inside the larger bubble.

4. Notice how the smaller bubble creates a compartment of air that is contained within but separate from the air inside the larger bubble.

d. Modelling membrane proteins

1. Lift the bubble frame out of the solution so that a thin film spans the frame. Hold the frame parallel to the tray.

2. Gently lay the small loop of thread on the film surface.

3. Use a pencil to break the bubble film INSIDE the thread's loop. The loop of thread should rapidly expand into the shape of a circle.

4. Insert your finger or a pencil into the middle of the thread loop.

5. Rock the frame repeatedly to see the thread loop drift across the film.

6. Repeat steps 1-5 with the larger string loop.

Discussion Questions:

- What properties of cell membranes could you model accurately with this activity?

- What properties of cell membranes were you unable to model accurately with this activity?

EARTH AND SPACE SCIENCE ACTIVITIES

Oreo Moon Phases

Context: Even though our students see the moon's phases every month, they don't often understand what causes them. In this activity, students will use Oreo biscuits to explore the concept of the moon's phases. This activity is designed to allow your students to visualise all of the moon's phases in a tasty way. Students can complete this activity in any year as it can be adjusted to make it age-appropriate.

Materials:

- Oreo biscuits

- Paddle pop stick or blunt knife
- Sheet of A3 paper
- Textas or coloured pencils

Instructions:

1. Research the phases of the moon and create a rough sketch of each phase.

2. Carefully split the Oreo cookies so that the cream is on one side and the other side is left 'clean'. Set aside the clean cookies for now.

3. Use a paddle pop stick to scrape the cream off their Oreo biscuits to make each phase.

4. Draw a picture of the Earth in the centre of the A3 paper and place the Oreo 'moons' in their appropriate order around the Earth.

5. Draw a Sun on the A3 paper in the appropriate place to create the various moon phases that are seen from Earth.

6. Label each of the phases of the moon and take a photo.

7. Enjoy the Oreos.

Discussion Questions:

- Describe the process that causes the moon to appear in these different phases.
- How long is one cycle of phases?

- Why does the same side of the Moon always face Earth?
- Will you see the same phases of the Moon in North America that you see in Australia?

Constellation Craft

Context: Astronomy can be difficult to teach with hands-on activities as we mostly see the students during the day. This activity allows students to research constellations and then use craft materials to recreate them in the classroom. This activity can be done in two ways. For example, firstly, the students could simply use black cardboard and stickers to create their constellation that is placed on the wall, or they could make a smaller version of their constellation that is placed over a torch in order to project the constellation onto the roof or against a window so that the light passes through the holes.

Materials:

- Thick, black cardboard
- Lead pencil
- Ruler
- Push pin
- Sticky tape
- Torch

Instructions:

1. Research constellations that are seen over your region and choose one to recreate.

2. Use the lead pencil and ruler to draw the constellation onto the cardboard.

3. Carefully pierce holes in the cardboard where the stars are located so that light is able to pass through them.

4. Place the cardboard over the torch or against the window to see the constellation.

Chocolate Fossils

Context: In this activity, students will use chocolate with small food items inside of them to explore the idea of fossils and extracting fossils. These chocolates can be pre-made so that students simply need to extract the fossils, or they can go through the process of melting the chocolate, adding their fossils, and placing them in a fridge to solidify themselves. This particular method involves using the pre-made chocolate and having the students explore the chocolates as 'palaeontologists'.

Materials:

- Chocolates with small food items inside of them (such as raisins, Smarties, M&Ms, or marshmallows)
- A tray or plate
- A spoon or spatula

Instructions:

1. Place the chocolates on the tray or plate.

2. Use the spoon or spatula to carefully extract the food items from the chocolates, taking care not to break or damage them.

3. Examine the food items and describe their characteristics (e.g. shape, size, colour).

Discussion Questions:

*   Explain how the process of making chocolate fossils is the same as what occurs when real fossils are made.

*   What is the difference between a mould fossil and an impression? Which type of fossil would the extracted chocolate fossils be classified as?

*   Discuss the challenges faced with extracting the fossils. How might this represent the challenges that palaeontologists face when carefully trying to remove fossils from the Earth?

Sweet Stratigraphy

Context: In this activity, students will use cupcakes made with different coloured layers of cake and lollies to explore the concept of stratigraphy and dating fossils in rocks. Stratigraphy is the study of the layers of sedimentary rock that make up the Earth's surface, and it is used to understand the history of the Earth and the evolution of life on our planet. By understanding stratigraphy, students can learn about the processes that shape the Earth's surface and the ways in which scientists use this information to study the past.

Materials:

- Cupcakes made with different coloured layers of cake and lollies
- Plate
- Knife
- Ruler
- Graph paper
- Coloured pencils

Instructions:

1. Give each group a cupcake and have them carefully cut it down in the middle.
2. Have the students examine the layers of the cupcake and describe their characteristics (e.g. width, colour, texture, ingredients).
3. Have the students use the ruler, graph paper, and coloured pencils to record the characteristics of each layer and the order in which they appear. Students should also make notes of any 'fossils' that are found within each layer.

Discussion Questions:

- Which natural process is the cupcake you examined today modelling?

- How do scientists use the fossils found in the Earth's layers to determine an area's age?

- Explain the concept of an index fossil and describe how it can be used to determine the age of other fossils that are found in an area.

Cookie Mining

Context: In this activity, students will use chocolate chip cookies to model the difference between open-cut and surface mining and their environmental impacts. Open-cut mining is a type of surface mining that involves removing layers of soil and rock to access deposits of minerals or other resources. Surface mining is a type of mining that involves removing the surface layer of soil and rock to access resources that are close to the surface. By understanding the differences between these two types of mining, students can learn about the environmental impacts of resource extraction and the importance of sustainable resource management.

Materials:

- Chocolate chip cookies

- A tray or plate

- A spoon or spatula

- A cutting board or other surface for chopping

Instructions:

- Place the chocolate chip cookies on the tray or plate.

- Using the spoon or spatula, demonstrate open-cut mining by carefully removing the top layer of the cookie to access the chocolate chips inside.
- Using the cutting board or other surface, demonstrate surface mining by carefully chopping the cookie into smaller pieces to access the chocolate chips.

Discussion Questions:

- Which of the two methods extracted the most chocolate chips?
- Discuss the differences between open-cut and surface mining and their respective impacts on the environment.
- Discuss ways that the environment could be restored in each scenario.

PHYSICS ACTIVITIES

Investigating Newton's First Law of Motion

Context: Inertia can be one of the most fun concepts to explore in physics. This can be done in a number of ways, including the 'tablecloth trick'. In this activity, students will use their understanding of inertia to get an egg into a cup of water without touching the egg itself. This activity is best carried out at the beginning of the unit on motion without giving the students tips or tricks on completing their challenge.

Materials:

- Cup or beaker half-filled with water
- A plastic plate or similar flat surface to sit over the top of the cup
- Toilet roll or cardboard tube
- Raw egg (this can also be done with a small ball, but the excitement of dropping a raw egg increases engagement!)

Instructions:

1. Show students a photo or drawing of the equipment set up as follows: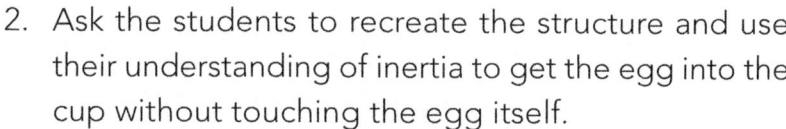

   a. Cup of water with the plate on top

   b. Toilet roll standing upright on the plate

   c. Egg sitting on the top of the toilet roll

2. Ask the students to recreate the structure and use their understanding of inertia to get the egg into the cup without touching the egg itself.

3. Once the students have achieved their goal, challenge them to stack their structure higher and repeat the process of getting the egg into the cup.

Discussion Questions:

- Using the language of Newton's first law, discuss why you were able to catch the egg in the glass when the plate and roll were hit.

- How lightly can you hit the tray and still have the egg fall into the glass? Is there a certain amount of force you need to apply?

- Does this work if swapping the egg for a much heavier or lighter object? How might you change how this experiment is conducted to explore Newton's second law of motion simultaneously?

Crash Test Dummies

Context - In a unit on forces and motion, students are required to explore Newton's Laws and their impacts on objects. In this particular activity, students will explore how Newton's first law of motion works in car accident scenarios and work to design a car safety system to protect their crash test dummies from experiencing extreme harm in an accident.

Materials:

- Dynamics trolley
- Ramp
- Ruler
- Chalk
- Solid barrier, e.g. brick
- Plasticine
- Talcum powder
- Sticky tape
- Paper and cardboard

- Scissors

Instructions:

1. Mould the shape of a person using the plasticine and lightly powder it to reduce stickiness.

2. Place the plasticine person on the dynamics trolley.

3. Set the ramp on a shallow slope and let the trolley run down it onto the floor. Note what happens to the plasticine person.

4. Place a chalk mark every 20 cm up the ramp and place a brick on the flat near the ramp's end.

5. Model a head-on collision by releasing the trolley from a 20cm mark on the ramp. Repeat for the rest of the marks. Note what happens to the plasticine person, particularly to any parts of the body that move a small amount and those that move significantly.

6. Determine which 20cm mark was the most 'life-threatening' to the plasticine person.

7. Build a sticky tape seat belt for the driver and repeat steps 3-6.

8. Take the belt off and use the paper and cardboard to add a 'crumple zone' to the front of the trolley.

9. Place the trolley and its driver on a flat desk. Model a rear-end collision by hitting the back of the trolley with your hand. Note what happens to the plasticine person, particularly to any parts of the body that move a small amount and those that move significantly.

10. Build a safety feature that would minimise injuries in this type of collision.

Discussion Questions:

- Which of Newton's three laws of motion is being demonstrated in this activity?

- Explain how the principles of force and energy transfer apply to car crashes and how this relates to the design and safety of cars and their occupants.

- How could this activity be modified to explore other aspects of vehicle safety and transportation engineering, such as the design and testing of crash barriers, the impact of road surface and weather conditions, or the use of advanced technologies such as autonomous driving and vehicle-to-vehicle communication?

Balloon Rockets

Context: Newton's Third Law of Motion states that 'for every action force, there is an equal and opposite reaction force.' Rockets use Newton's Third Law to move. An action force is produced when burning fuel accelerates out of the back of the rocket. The reaction force of pushing out the fuel is called the thrust. This force makes the rocket accelerate away from the fuel. In this activity, students will use a balloon to model the way that rockets take advantage of Newton's Third Law of Motion.

Materials:

- Balloon
- String
- Tape
- A drinking straw
- Scissors
- Video camera

Method:

1. Tie or tape one end of the string to a secure surface or object.

2. Trim the straw to a length of 2 - 5 cm and thread the drinking straw onto the string.

3. Tie or tape the other end of the string to another surface or object, ensuring that the string is horizontal and taut.

4. Blow up the balloon and hold the end shut. Do not tie the balloon off.

5. Tape the balloon to the drinking straw while still holding the end of the balloon shut.

6. Pull the balloon to the end of the string – make sure that the open end of the balloon is facing you.

7. Release the balloon and record your observations.

8. Repeat steps three more times and make sure to video record one of the tests.

Discussion:

- Describe the force acting on the balloon that caused it to accelerate along the string. Think about the action force and reaction force that occurred due to Newton's Third Law.

- The four forces acting on the balloon immediately after it was released were thrust force (also called reaction force), friction, gravitational force and support force. Discuss these forces and how each of these forces works to make a rocket blast off.

- In this experiment, the balloon may have stopped moving before reaching the string's end. Explain why the balloon stopped moving.

Fruity Circuits

Context: In this activity, students will create electrical circuits using fruit and compare the voltage produced by each type of fruit. By understanding the principles of electricity and the properties of different fruits, students can learn about the relationship between electricity and chemical reactions.

Materials:

- Various types of fruit (e.g. lemon, potato, apple, orange)

- A multimeter or other device for measuring voltage

- Wires with alligator clips

- Graph paper or other materials for recording data

Instructions:

1. Using the wires and alligator clips, create electrical circuits using the different types of fruit.

2. Using the multimeter or other device, measure the voltage produced by each circuit.

3. Record the voltage produced by each fruit on graph paper or other materials.

4. Compare the voltage produced by each fruit and discuss the results.

Printed in Great Britain
by Amazon